CALIFORNIA NATURAL HISTORY GUIDES

INTRODUCTION TO EARTH, SOIL, AND LAND IN CALIFORNIA

California Natural History Guides

Phyllis M. Faber and Bruce M. Pavlik, General Editors

Introduction to

EARTH, SOIL, AND LAND IN CALIFORNIA

David Carle

UNIVERSITY OF CALIFORNIA PRESS

Berkeley   Los Angeles   London

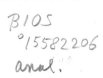

This book is dedicated to Janet and our sons, Nick and Ryan.

University of California Press, one of the most distinguished university presses in the United States, enriches lives around the world by advancing scholarship in the humanities, social sciences, and natural sciences. Its activities are supported by the UC Press Foundation and by philanthropic contributions from individuals and institutions. For more information, visit www.ucpress.edu.

California Natural History Guide Series No. 101

University of California Press
Berkeley and Los Angeles, California

University of California Press, Ltd.
London, England

Library of Congress Cataloging-in-Publication Data

Carle, David, 1950–
   Introduction to earth, soil, and land in California / David Carle.
     p. cm. — (California natural history guide series; no. 101)
   Includes bibliographical references and index.
   ISBN 978-0-520-25828-0 (cloth: alk. paper)
   ISBN 978-0-520-26681-0 (pbk. : alk. paper)
   1. Soil ecology—California. 2. Soil animals—California.
3. Land use—California. I. Title.

QH105.C2C366 2010

508.794—dc22   2009050518

Manufactured in China
19 18 17 16 15 14 13 12 11 10
10 9 8 7 6 5 4 3 2 1

The paper used in this publication meets the minimum requirements of ANSI/ NISO Z39.48-1992 (R 1997) (*Permanence of Paper*). ∞

Cover photograph: Decomposition recycles the nutrients in a fallen tree, enriching the soil and fertilizing new growth. Photo by David Carle. All unattributed photographs captioned on p. 209.

The publisher gratefully acknowledges the generous
contributions to this book provided by

the Gordan and Betty Moore Fund
in Environmental Studies
and
the General Endowment Fund of the
University of California Press Foundation.

CONTENTS

ACKNOWLEDGMENTS

My thanks to Rick Kattelmann and Sally Gaines for reading a draft of this book, a favor they also performed for the earlier titles in this series, which have dealt with water, air, fire, and now, earth in California. Stan Bluhm (with Coastwalk California), David Runsten (program director for the Community Alliance with Family Farmers), Scott Crosier (geography professor at Cosumnes River College), and Jeff Lustig (professor of government at CSU–Sacramento) also read this text.

The image search led me to these particularly helpful people, to whom I am very grateful: Eddie Dunbar, with the Insect Science Museum of California (www.bugpeople.org); Peter J. Bryant, School of Biological Sciences, UC–Irvine (http://nathistoc.bio.uci.edu/); John Louth, with the USDA Forest Service, White Mountain Ranger District; Dace Taube, at the University of Southern California digital archives; Mike Darnell, with American Farmland Trust; Theresa Kiehn, at Great Valley Center; Kerry Arroues, soil scientist with the USDA Natural Resources Conservation Service; Oksana Gieseman, for the Soil and Water Conservation Society; Mike Lynch, for California State Parks history and photographs; Fred Andrews, State Park Interpreter in the Mendocino District; Nancy Bailey, at Lassen National Park; and Dr. Michael Amaranthus, of Micorrhizal Applications, Inc., and Frans Janssens, from the University of Antwerp, Belgium.

Phyllis M. Faber, a general editor of the California Natural History Guide series, got me started on this series of books years ago by asking for "a simple primer" on water issues. Her special interest in a book dealing with the living soil inspired and shaped the present effort. The other general editor for the series, Bruce M. Pavlik, kept me motivated with advice and philosophical thoughts about humanity's relationship with the land.

The editors at UC Press have been wonderful to work with on all four books. Thank you again, Jenny Wapner and Kate Hoffman, for your support and expertise.

I also want to acknowledge and thank my longtime literary agent, Julie Popkin.

Finally, I am ever grateful to my first reader, Janet, for her willingness to accompany me on quests for photographs and information and her endless patience with my hours at the computer.

INTRODUCTION

How can I stand on the ground every day and not feel its power? How can I live my life stepping on this stuff and not wonder at it?

WILLIAM BRYANT LOGAN (1995, 2)

Stand unshod upon it, for the ground is holy, being even as it came from the Creator. Keep it, guard it, care for it, for it keeps men, guards men, cares for men. Destroy it and man is destroyed.

ALAN PATON (1987 [1948], 33)

When it rained, earthworms would mysteriously appear on the sidewalks. The first seven years of my life were spent in southern California, on a street with sidewalks and rows of houses, each with a small yard. Because rain was limited to a few winter months, those waterlogged worms were rarely seen. Their emergence always came as a surprise. In those early years, before the invention of so many electronic toys, we spent lots of time outdoors closely communing with dirt. We dug in the bare patches of our yards with kid-size shovels and buckets, making roads for toy trucks and earthen walls for toy soldiers. We inevitably exposed pill bugs, beetles, cutworms, and earthworms, and occasionally intersected larger holes tunneled by moles or gophers. Those excavations also revealed small roots, mysteriously emerging from out beyond the edges of our holes. Our patient mothers had to deal with endlessly dirty clothing, but their endlessly dirty children were learning lessons about living soil.

Most of the earth in our yards was covered by crabgrass lawns, with trees and shrubs around the edges. A few neighbors eventually converted to "low-maintenance" yards—nothing but gravel, ornamental rocks, and cement—unbearably hot places in the summer and good for playing on only if they provided smooth surfaces for roller skates, bikes, or basketballs. Growing up in Orange County during the 1950s, we also watched as bulldozers took out the orchards that we once bicycled past on our way to school. We saw houses inexorably replacing farmland. New roads and new construction required that rich topsoil be graded, scraped away, and paved over in those southern California counties, which had led the nation in agricultural productivity through the first half of the twentieth century.

Only when earthworms reappeared on the sidewalks during the rain did I stop to wonder what happened to all of the living things underground after they were covered over by buildings and pavement (pl. 1).

Introduction to Earth, Soil, and Land in California begins with the story of ecosystems beneath our feet and the living soil. This is my fourth, culminating book in the series Californians and Their Environment, following books about the state's water, air, and fire. "Earth" is the logical next subject, but the concepts presented here are less about geology than about environmental geography; less about rocks, faults, and tectonic plates than about the life that transforms inorganic "dirt" into "living soil," and the land itself as the fundamental substrate of our terrestrial environment. The goal is to help readers understand humanity's ultimate reliance on the marvels of living soil and Earth's natural cycling processes as sources of life.

A living soil forms when inorganic dirt (broken down rocks and minerals) is integrated with organic plant and animal matter. Incredibly diverse and sizable populations

Plate 1.
When it rains,
earthworms
may emerge
from the
ground.

of bacteria, fungi, insects, worms, and other life forms are found in fertile soils. Such living soils are the basis for almost every terrestrial ecosystem.

Water and mineral cycles send nutrients and chemicals through the atmosphere and back to the soil. Soil is the ultimate place of connection, nurturing growth while life continues and then recycling its components after death. Earth's soil and landforms become habitat for vegetation and wildlife, become biomes shaped by parent rock, topography, elevation, exposure, climate, and time to form the specific conditions for communities of life.

The opening section of the book is a basic primer on soil ecology that introduces the processes and communities of organisms that build and replenish soil. Profiles and categories scientists use to describe soil variations are introduced, along with a few of the fascinating soil organisms that spend part or all of their life cycles out of sight and out of mind below ground.

Repeating a progression used in the earlier three books in the series, elemental concepts lead to broader topics in an overview of humans on the land in California. "The Human Footprint on California Land" explores Native American attitudes toward land before there was a state called "California." Land ownership transitions followed with the Spanish and Mexican grazing economy, the Gold Rush, and early statehood, when land became concentrated in strikingly few hands as homestead and other land grant acts were manipulated. Establishing the boundaries that give California its unique shape involved contentious debates and some fascinating history. Latitude and longitude lines and the alternative mapping grid that produced townships, ranges, and sections helped turn "land" into "real estate."

"Wild Land" considers how wildland management relates to soil health, including the role of fire in soil ecosystems. Habitat loss and competitive pressure from human population growth explain the great loss of biodiversity and the ever-lengthening list of endangered species in this state. Increasingly valuable wild and open-space lands are being managed by a complex assortment of local, regional, state and federal agencies, and nongovernmental organizations.

California also remains an "Untamed Land," notorious for its unsettling number of earthquakes, wildfires, landslides, and less frequent volcanic events. The human risks and societal costs of life on the California landscape, along with the widespread assumption that "it won't happen in my lifetime" run up against "Reality Checks."

"The Living Soil Made My Lunch" considers not only the ways in which farms depend on soil fertility and influence carbon sequestration, but also problems of groundwater overdrafting, soil degradation, land subsidence, and salinity increases in the root zones of irrigated soils. "Death by a Thousand Cuts" looks at the seemingly unending urban growth pressures that consume prime farmland, along with efforts to protect agricultural land and productivity.

Finally, "Walking Softly" explores "California's Footprint on the Earth": the state's impacts on the planet. The ecological footprint concept calculates how many acres of global land it takes to support each person, to learn how well that estimate conforms with a sustainable future. The value of this exercise is to help Californians to recognize choices they can make as individuals to minimize their ecological and carbon footprints. From landfills to "smart growth" choices, in a state with 38 million residents and an economic model that has yet to turn away from everlasting growth toward long-term stability, the scale of California's impacts creates challenges and tremendous problem-solving opportunities. "Compassionate Numbers" takes a direct look at the problems of population growth in California that are too often avoided, and their influence on just about every other challenge and topic addressed here, from earthworms to earthquakes to Earth Day.

The goals of this book are to turn "people and the land" concepts back toward their basis in the earth, to encourage people to appreciate the living soil as the literal foundation for environmental concerns, and to help them understand soil connections with the water, air, and fire topics covered in the other books in the Californians and Their Environment series.

As you read, consider the earth directly below—however many layers of chair and floor and foundation lie between you and the ground. Once it was parent rock, then mineral "dirt" that became "soil" when life was added. The first humans in California may have considered it a portion of land they called home, though they probably never claimed it personally as their own. Later, newcomers saw the region's land in a new way; they "took title" and introduced grazing animals and exotic grasses. Others followed with different intentions about the best use of the land. It became "real estate," part of a land parcel that could be split among owners for a price. New lines delineated that location on Earth: within a state called California; part of a county, and perhaps within a city's limits; covered now by a house, school, or business building, or perhaps simply landscaped with garden and lawn.

It is not easy to understand and value those living things that are invisible and mostly remain hidden below ground. Yet such an appreciation by modern Californians, whose lives often seem distant from the earth, is the aim of this book, because the living soil remains the fundamental basis of life.

Soil isn't a granular medium suffused with chemicals, it is alive, and must be alive to function.

GARY JONES (2005)

The nation that destroys its soil destroys itself.

FRANKLIN DELANO ROOSEVELT, IN A LETTER
TO ALL STATE GOVERNORS ON A UNIFORM SOIL
CONSERVATION LAW, 1937 (WOOLLEY AND PETERS, 2010)

SOIL IS MORE than "dirt," and land is far more than "real estate." The distinctions lie with the organisms that teem in the earth beneath our feet and the processes that transform that earth into living soil.

The ground physically supports us, our structures, and everything else on land, but soil is life's foundation in many other ways. Soil sustains everything alive by producing food and by exchanging water vapor and oxygen and nitrogen gases with the atmosphere. When an organism dies, decomposers in the earth recycle the body, and the products of this decomposition enable new life to emerge. Soils store and cycle nutrients, including carbon, nitrogen, phosphorus, and minerals that fertilize plant growth. Soil also filters and detoxifies wastes and pollutants.

Soil is a mix of minerals, organic matter, water, and—perhaps surprisingly, given the apparent solidity of soil—air. Air percolates down to reach respiring root cells and other living organisms that depend on it. Oxygen moves through pores and channels between soil particles, which also pass carbon dioxide back up into the atmosphere. Every hour or so all of the air in the top eight inches of soil is exchanged (unless the soil is saturated with water).

The groundwater surrounding soil particles can be slowly cleansed of dissolved contaminants by soil microorganisms. The depth at which water completely saturates all underground pores and channels is called the "water table." This is the underground reservoir sought by well diggers. Stored groundwater eventually finds its way back to the surface and supplies springs, rivers, and lakes, rejoining the above-ground water cycle.

Soil Profiles and Categories

Soil formation begins with inorganic parent material, which is altered over time, worked on by climate, and processed by topography and living organisms. The parent material is the bedrock underlying the layers of soil. Bedrock is a product of volcanic eruptions, sediments that filled ancient seas and later were uplifted into new land forms, or collisions of the tectonic plates that form Earth's crust. When one tectonic plate is forced beneath another, a phenomenon known as subduction, the resulting heat and pressure can transform rock into new forms. The result is bedrock, with a unique structure and chemistry that translates into unique soil types. Climate also works on bedrock. Freeze and thaw cycles cause the rock to expand and contract, and this breaks it into smaller pieces. Rainwater leaches chemicals, and temperature fluctuations alter the rates of chemical reactions in the ground (pl. 2).

Parent material can also be dissolved by the acids secreted by lichens growing on rock. The tiny openings that result allow water to enter, so that freeze-thaw expansion and contraction speed the rock's breakdown. Lichens are formed by two cooperating organisms: fungi

Plate 2. Water and gravity break down rock to produce soil at the base of this hill in Pinnacles National Monument.

Plate 3. A colorful splash shows where lichens are secreting acid in the initial phases of soil building.

and algae. Fungi provide the outer body and anchor the lichen to rock. Algae live inside the fungal cells and carry on photosynthesis, feeding themselves and their fungal hosts in a symbiotic relationship (pl. 3). The growth of plant roots and the burrowing of animals can also break parent rock apart.

Elements, the chemical building blocks listed on the chart found in most science classrooms, combine to form minerals. Ninety-eight percent of Earth's crust is made of just eight elements: oxygen, silicon, aluminum, iron, calcium, sodium, potassium, and magnesium.

California occupies about 100 million acres of land. The diversity of this state's landforms—including mountain ranges, valleys, and desert expanses—explain much of its high soil diversity (map 1). California stretches from north to south for one thousand miles, spanning 10 degrees of latitude. Mountain spines—the Sierra Nevada, Coast Ranges, and Transverse Ranges—outline coastal and inland valleys. California has the lowest elevation of the 48 contiguous states,

Map 1. Landform provinces in California.

found at Death Valley (282 feet below sea level), and, not far away, the highest point, the summit of Mount Whitney (14,494 feet above sea level). This topographic diversity combines with weather patterns to shape local climates, break down rock material, and produce a variety of soils.

The ups and downs of a landscape's topography dictate erosion rates, as water and gravity break up rock and move the particles, eventually depositing them as sediments. Most of the state's precipitation comes from the west, delivered by winter storms arriving from the Pacific Ocean. Moist air moves inland, pushed by the prevailing west winds, and rises over the state's mountain backbones. The uplifted air cools and loses its hold on water vapor, and this is why the western slopes of the state's mountains receive the heaviest precipitation. As moving air descends the eastern slopes, it is warmed by compression and releases less moisture; as a result, dry "rain shadows" characterize the desert regions downwind from the major mountain ranges. The force of water moving across the landscape sculpts the earth, forming features that range from tiny rivulet channels up through the V-shaped river canyons that drain mountain watersheds. Glaciers, advancing and retreating in response to long-term climate variations, scraped away topsoil and carved the U-shaped valleys of the Sierra Nevada.

Wind modifies soil by sorting matter, leaving behind heavier material and carrying lighter particles aloft. Dunes form in the desert and along coastal beaches, where the wind speed slows enough to allow blowing sand particles to drop to the ground.

Soils that have formed in place, above parent bedrock, are called residual soils. This category represents over half of California's land, including forested upland locations (map 2). Transported soils, by contrast, result when water and wind shift material downhill or downwind. These forces, water and wind, created California's primary agricultural soils in its valleys and basins, and on terraces along the edges

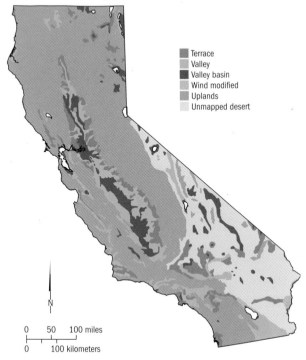

Map 2. Soil terrain categories in California.

Terrace
Valley
Valley basin
Wind modified
Uplands
Unmapped desert

0   50   100 miles
0   100 kilometers

of valleys or on coastal plains. Layers of rich alluvium can build up in bottomlands. Glaciers grind bedrock into fine "till," which is transported downstream by meltwater and adds to lowland deposition. Wetland soils occur along the edges of water channels, where plant roots find a constant source of moisture, or across broader parts of the landscape where fine clays or underlying hardpan slow drainage, keeping groundwater close to the surface.

The chemical and physical features of parent rock greatly influence the pattern of <u>downstream</u> or <u>downwind</u> deposits. The saline soils of closed desert basins such as Death Valley

Plate 4. The closed basin of Death Valley accumulates salts in the desert soil that crystallize as water evaporates.

are one example (pl. 4). Another is the unusual water chemistry of Mono Lake, where the salinity of exposed lake beds is the result of dissolved minerals brought by streams from the surrounding uplands into a basin with no water outlet. A unique blend of chlorides, carbonates, and sulfates is the product of a watershed formed under the influence of volcanic action and the weathering of the granitic Sierra Nevada rocks.

Most of us can recognize only very obvious soil variations—whether a piece of land is rocky, sandy, or saturated with water, for example. Boundaries between plant communities provide clearer signals that the variations we observe above ground are shaped by soil characteristics. "Edaphic" species are found where soil conditions are toxic to other plant species. Those conditions become the primary factor, rather than moisture or topography, that determines what plant species can grow in such areas. For example, the ancient bristlecone pines *(Pinus longaeva)* thrive in the

Plate 5. Bristlecone pines thrive in the alkaline dolomite soil of the White Mountains where most other plants are excluded. A sharp delineation where sagebrush scrub becomes the dominant vegetation corresponds with sandstone soils.

White Mountains near the eastern boundary of the state, primarily on soil containing dolomite, a mineral whose whitish appearance gives the mountain range its name. The high alkalinity of dolomitic soil excludes most other plants, giving the bristlecones a competition-free habitat. Abrupt vegetation boundaries are visible where the bristlecone forest gives way to sagebrush (*Artemisia* spp.) shrubs, which prefer sandstone soils (pl. 5).

Soil formation is a very slow process, but the reverse—loss of soil through wind and water erosion—can, unfortunately, proceed quickly. As much as 1 percent of Earth's topsoil—the most alive, organism-rich soil layer—is eroded each year, washed or blown away, a process that gradually reduces the productivity of the land. Civilizations have failed after depleting their farm soils through plowing and other cultivation-related disturbances that exposed the ground to forces of erosion, or by growing crops that consumed nutrients faster than they could be replenished.

Societies have declined or disappeared when populations were forced to migrate after their soils had lost their fertility. Writing about the history of soil depletion in the Middle East region once known as the Fertile Crescent, David Montgomery has pointed out how "the televised images of the sandblasted terrain of modern Iraq just don't square with our notion of the region as the cradle of civilization" (Montgomery, 2007, 4). John Steinbeck's novel *The Grapes of Wrath* explored the Dust Bowl calamity that devastated the western plains states and led to a mass migration to California that shaped the state's population and agricultural labor issues during the 1930s (pl. 6).

Soil scientists call the visible layers in a soil profile "horizons" and designate them with a letter: O, A, E, B, or C. The O horizon is the organic layer on the surface. It is made of recently deposited plant and animal debris and partially decomposed matter. The A horizon, also called "topsoil," is the layer near the surface that has much more organic

Plate 6. Dust Bowl conditions on the nation's central plains prompted new concerns about topsoil conservation and a mass immigration to California in the 1930s.

matter than can be found in the lower layers. The A and O horizons contain most of the plant roots and soil organisms in a given profile. The E horizon comes next (though it is alphabetically out of order); it designates the layer altered by weathering, where water has dissolved minerals such as iron and aluminum and moved them and organic matter downward. The result often is a change in color where those products have settled into the B horizon. Finally, the C horizon consists of the parent material below the other horizons (pl. 7, fig. 1). Exceptions to this tidy pattern of layering occur where soil has been shifted away from its source rock by floods and glaciers or dropped onto a site by wind or volcanic eruption.

Texture is an important characteristic of soil, determining how readily water percolates, how easily roots grow, and how much air can move in the spaces between soil particles. Texture, a function of particle size, is defined by three extremes:

Plate 7. Dark topsoil distinguishes the "O," or organic, layer from the A horizon of the topsoil, where most living soil processes are concentrated.

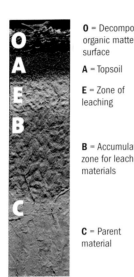
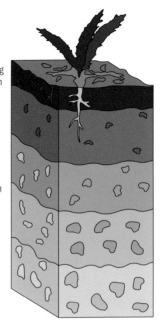

O = Decomposing organic matter on surface

A = Topsoil

E = Zone of leaching

B = Accumulation zone for leached materials

C = Parent material

Figure 1. Soil horizons.

sand, silt, and clay. Sand particles are the largest; they can be seen with the naked eye. A microscope is needed to distinguish silt particles, which have the feel of flour when rubbed between the fingers. Clay particles are extremely fine, flat particles that stack tightly together and therefore feel smooth and slippery. They have a negative electrical charge that enables them to attract and hold many positively charged plant nutrient ions, such as potassium, calcium, and ammonium ions. Negatively charged nitrate ions, by contrast, are easily washed from clay soil, and as a result low nitrogen levels often limit plant growth in clay.

A soil may simply be one of those texture types—for example, deserts generally have sandy soils—but the categories also blend in proportions that generate a dozen descriptors:

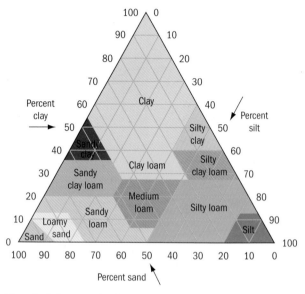

Figure 2. Soil texture triangle.

"clay loam," "loamy sand," and "sandy loam," to give a few examples. The full range of combinations is often depicted using a soil texture triangle (fig. 2). Loam is a blend of the three primary particle types and is the soil type preferred by gardeners and farmers. As shown on the triangle, a mix of 40 percent sand, 40 percent silt, and 20 percent clay results in a medium loam. That mix of textures holds on to soil moisture while still allowing air, water, and roots to penetrate.

Add organic matter to loam, and the fertile result is humus. Dark with an earthy, wholesome smell, humus is full of organic compounds that further help to bind soil particles together to resist erosion and hold water (pl. 8).

Formal soil classifications are based on organic content, moisture level, age, weathering, and oxidation. Twelve orders cover all of the world's soil types. The terminology is unfamiliar to most of us, but characteristic descriptions relate to

Plate 8. Humus, a fertile mix of organic and mineral matter, has a pleasant earthy smell that is produced by actinomycetes bacteria.

familiar landforms and sections of the state. For example, *aridisols* are the dry soils of deserts and the southern San Joaquin Valley, where salts and calcium carbonate may accumulate near the surface that would leach away in wetter environments. *Mollisols* are the soft, fertile soils of valley grasslands and foothill savannas (mixed grass and trees). Rice fields in the Sacramento Valley are found on *vertisols,* which are clayey soils that do not drain well and are subject to vertical swelling, shrinking, and cracking when moisture levels change. Soils that contain aluminum (Al) and iron (Fe) washed down from higher land are called *alfisols* and characterize foothill chaparral and oak woodland plant communities.

Inceptisols are newer soils that have not developed much beyond the parent material; these are found on the upper reaches of the Sierra Nevada and the Klamath Mountains. Farther down on the slopes of mountains, soils are older, and extremely weathered *ultisols* are associated with coniferous forests. On floodplains that regularly receive deposits of alluvial sediments washed from higher lands, or on slopes that are actively eroding, *entisols* represent the undeveloped condition where no diagnostic soil horizons are visible. *Histosols* are the dark peat soils found in the Sacramento–San Joaquin delta wetlands, where organic material accumulates faster than it can be decomposed. California's northeastern Modoc Plateau has dark *andisols*, soils formed by volcanic ash.

Within orders, soils are further classified into series generally named for a specific location where a characteristic example of the type is found (referred to as the soil's type location). Over a thousand series have been described within California (there are 23,000 soil series in the United States!). Things become very specific and complex at this level. One example is the Yolo series, a fine, silty, alluvial loam that is valuable for farming. Its type location is on the campus of the University of California at Davis in Yolo County (pl. 9).

In Orange County, where my childhood years were spent, a number of soil series exist, but Myford Sandy Loam was "my" dirt to play in. Its type location is about 1.2 miles east of the junction of San Diego and Santa Ana freeways on the

Plate 9. The Yolo soil series is the characteristic soil type farmed near Davis, California.

Irvine Ranch. Much of that fertile soil, which allowed that region to become a leading citrus producer for the nation in the early twentieth century, has been sealed beneath pavement and concrete.

Each state in the Union has a designated state soil. California's is the San Joaquin series, which corresponds to 500,000 acres of fertile farmland in the Central Valley. It is a sandy loam deposited by Sierra Nevada streams where they spread out onto alluvial fans during the glacial periods. Reddish material at the surface overlies increasing clay content and, below the productive layers, a cemented hardpan that limits root and water penetration. The type location is north of Lodi in San Joaquin County (pl. 10).

Plate 10. The San Joaquin soil series is California's official state soil. Below the fertile topsoil layers is a hardpan that limits water movement and root penetration.

California also has an official state rock, serpentine, which is a dull green or sometimes bluish rock with a mottled appearance that is suggestive of a serpent's skin. Serpentine rocks feel greasy or soapy. Soils derived from serpentine bedrock appear gray or white. Serpentine soils correspond to edaphic plant and animal species found only on these soils. One can encounter abrupt changes from dense forest or grassland to barren or thinly vegetated areas when crossing boundaries onto these soils. Serpentine soils are deficient in necessary plant nutrients such as calcium, nitrogen, phosphorus, potassium, and the micronutrient molybdenum. They do, however, contain iron or magnesium silicate and heavy metals such as nickel and chromium that are toxic for most plants. Arthur R. Kruckeberg, Professor of Biology at the University of Washington, called serpentine soils one of many "kooky soil" types of California in his book *Introduction to California Soils and Plants* (map 3).

Nearly 285 plant species or subspecies are restricted to serpentine soils, including 20 percent of the state's rare endemics. Leather oak *(Quercus durata* var. *durata)*, is found almost exclusively on central and northern California serpentine soils and, along with whiteleaf manzanita *(Arctostaphylos viscida)*, can dominate a low-growing chaparral shrub plant community. Isolated trees that may appear among the shrubs include the Sargent cypress *(Cupressus sargentii)* and gray pine *(Pinus sabiniana)*. Almost no plants grow in serpentine barrens in places such as San Benito, in the inner Coast Ranges (pl. 11).

Dust blown off serpentine soils may carry asbestos fibers—an air quality concern, because inhaling the tiny fibers can cause lung disease. Serpentine rock is formed at tectonic subduction zones such as those near the coast of California, where the Pacific plate is being forced beneath the North American plate. There heat and pressure can transform peridotite rock, the most common rock in Earth's mantle, into serpentine. This soil type is also found along the

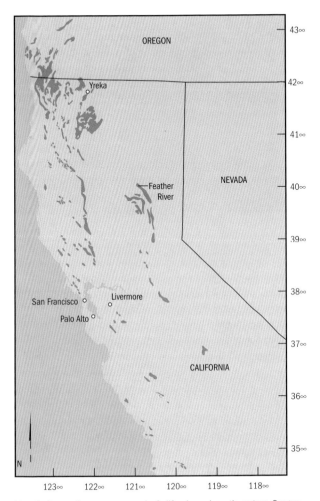

Map 3. Serpentine occurrences in California and southwestern Oregon.

west slopes of the Sierra Nevada, associated with the tectonic forces that lifted those mountains.

In addition to an official state soil (the San Joaquin series) and an official state rock (serpentine), California has an official

Plate 11. Serpentine barrens and serpentine chaparral in Lake County.

state mineral—gold. A famously nonreactive metal, gold plays only a minor role in soil ecosystems, although the excitement it generated among humans back in the nineteenth century shaped events and attitudes about the land that explain much about modern California. The story of those events will unfold, but the natural history of organisms within the living soil should come first.

Cities in the Soil

> *Our knowledge of ecosystems is fundamentally distorted by our above-ground, visual perception of nature and our ignorance of life below-ground.*
>
> BOUCHE (LEE, 1985, IX)

> *The entire ground habitat is alive. Living forms create virtually all of the substances that flow around the inert grains.*
>
> EDWARD O. WILSON (2010, 70)

Everything we see above ground is tied to essential processes going on beneath the surface, generally out of our sight and beyond our awareness. Almost half of the volume in most topsoil is dirt or mineral matter; the rest is mostly air or water. Living organisms and their recycled remains may account for only 5 percent of the soil's volume, yet they are the essential modifiers that transform "dirt" into living ecosystems. The rich world of microbes, plants, animals, and minerals is a complex, busy "city in the soil."

Soil food web diagrams simplify the incredibly complex pathways used in nutrient cycling (fig. 3). The sun's energy

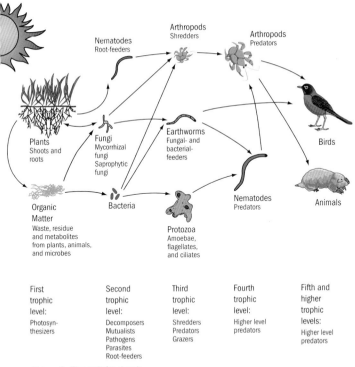

Arthropods
Shredders

Arthropods
Predators

Nematodes
Root-feeders

Plants
Shoots and
roots

Fungi
Mycorrhizal
fungi
Saprophytic
fungi

Earthworms
Fungal- and
bacterial-
feeders

Birds

Organic
Matter
Waste, residue
and metabolites
from plants, animals,
and microbes

Bacteria

Nematodes
Predators

Animals

Protozoa
Amoebae,
flagellates,
and ciliates

First trophic level:	Second trophic level:	Third trophic level:	Fourth trophic level:	Fifth and higher trophic levels:
Photosyn- thesizers	Decomposers Mutualists Pathogens Parasites Root-feeders	Shredders Predators Grazers	Higher level predators	Higher level predators

Figure 3. The soil food web.

sets the food chain in motion. Solar energy is captured through photosynthesis by plants, which use this energy to build complex carbon molecules in their stems, leaves, and flowers above ground and their root systems below ground. These roots, plus the wastes and the decomposing remains of organisms, nourish bacteria, fungi, single-celled protozoans, nematodes (tiny roundworms), earthworms, insects, and other arthropods, plus mammals, reptiles, and birds that forage in the ground. Plant roots modify the environment immediately surrounding them, a region called the "rhizosphere," by exuding chemicals that actively attract and then feed microbes and fungi. These close associates benefit the plants by capturing water, minerals, and nutrients from the soil.

Along food chains, organic compounds and minerals pass from plants to herbivores to predatory carnivores. Bacteria are eaten by larger microbes such as amoebae, paramecia, and other single-celled protozoans. Fungi- and plant-eating nematodes are food for mites, insects, and spiders. Some predacious nematodes consume bacteria, fungi, and tinier arthropods. Earthworms are generalists, feeding on detritus and on soil and plant residues but they also eat microbes. Generally, macroorganisms prey on microorganisms and, in turn, provide food for even larger animals.

Everything that lives in or on the soil generates metabolic wastes. These wastes, along with the chemicals produced by decomposition following death, enter recycling processes that furnish nitrogen, phosphorus, and other essential fertilizing minerals to plants. Soils may simply retain the carbon from decomposed plants and animals for long periods. The world's soils store four times as much carbon as the atmosphere and three times as much as all Earth's trees.

Soil Respiration: The Carbon Cycle

Soil organic matter is approximately 50 percent carbon, the element that provides the framework for all organic molecules, the chemical building blocks of life. Atmospheric carbon enters land ecosystems when carbon dioxide is captured through photosynthesis by green plants or cyanobacteria (fig. 4). Photosynthesis (literally, "making with light") uses solar energy to transform carbon dioxide and water into glucose, which is used by plants to build complex carbon molecules. Oxygen gas is released into the atmosphere during photosynthesis. This reaction is summarized as: carbon dioxide + water + sunlight (energy) → sugars + oxygen. Aerobic respiration reverses the equation: it breaks apart complex carbon molecules in the presence of oxygen and releases carbon dioxide, water, and energy.

Earth's topsoil stores 1,500 gigatons of carbon, according to the United Nations Food and Agriculture Organization. That compares with 650 gigatons of carbon held in the planet's vegetation and 750 gigatons in the atmosphere. The soil's organic matter serves as the largest carbon reservoir that interacts with air. This relationship is being closely studied as the threat of global warming looms because of the elevated levels of carbon dioxide (an atmospheric greenhouse gas) released since the onset of the Industrial Revolution.

The increase in atmospheric carbon dioxide is due mostly to the burning of fossil fuels—ancient carbon deposits stored in the ground. However, more than one-third of the carbon dioxide added to the atmosphere in the last 150 years has been tied to land use; over that time, the world has lost 50 to 80 percent of its topsoil. When soil is plowed and turned by tilling, the organic material in the soil is exposed to oxidation and to wind and water erosion. Carbon sequestration, or long-term storage within the soil, is an approach being considered for reducing the amount of carbon dioxide released into the atmosphere. As an added benefit, soils with more organic carbon are more fertile and productive (see "Dirt First!").

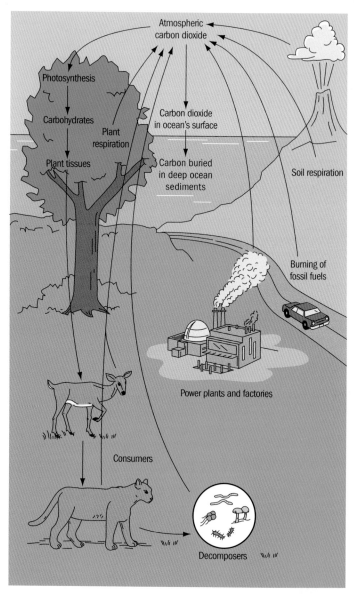

Figure 4. The carbon cycle.

Soil Respiration: The Nitrogen Cycle

Nitrogen is a key element in proteins, enzymes, vitamins, and DNA molecules. Although nitrogen gas (N_2) makes up about 78 percent of the air, most living organisms cannot make direct use of this nonreactive molecule. Energy is necessary to "fix" nitrogen by chemically converting it to nitrate (NO_3) or ammonia (NH_3), forms that can be used by plants and passed along to feeding animals (fig. 5). The major nitrogen fixers are bacteria that live in the soil or in specialized root cells. Of the tens of thousands of bacteria species, perhaps fewer than 200 are free-living nitrogen fixers, residing outside of roots. Nitrogen-fixing rhizobia, however, are bacteria housed in root nodules, where, in a symbiotic relationship with the host (legume plants, such as peas, beans, and clovers), they

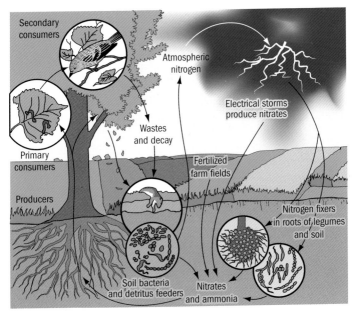

Figure 5. The nitrogen cycle.

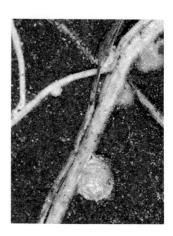

Plate 12. Roots of legume plants have nodules that house nitrogen-fixing bacteria.

share nitrogen compounds in return for the energy-rich carbohydrates produced by photosynthesis (pl. 12). Mesquite shrub roots also house nitrogen-fixing bacteria that help maintain nitrogen levels in the nutrient-deficient soils of the Mojave Desert. On most soil food web diagrams, species that form such close, mutually beneficial relationships are designated "mutualists."

The cycling of nitrogen is completed with the excretion by animals of urea or the breakdown of amino acids (the building blocks of proteins) during decomposition. Denitrifying bacteria (common in wetland muds and estuaries) convert amino acids to nitrogen gas, which then returns to the atmosphere.

Nitrogen is often a limiting nutrient for plants. Nitrogen levels drop because nitrates are water-soluble and are easily washed away. Farmers counter this problem by planting legumes in rotation with other crops as a "green manure" to replenish the nitrogen in the soil. They also apply ammonia and nitrate as chemical fertilizers. Since World War II, the Haber-Bosch process has been used to manufacture such fertilizers, using heat and pressure to combine nitrogen and hydrogen in the presence of a catalyst. The process requires considerable energy. Once commercially manufactured nitrogen fertilizers became available, "instead of eating exclusively from the sun, humanity . . . began to sip petroleum," wrote Michael Pollan in *The Omnivore's Dilemma* (2006, 45). However, the acidity of such chemical fertilizers can damage organisms that are essential to the soil ecosystem and create lifeless soils, totally dependent on the application of manufactured fertilizer.

Decomposition

> *I bequeath myself to the dirt, to grow from the grass I love;*
> *If you want me again, look for me under your boot-soles.*
>
> WALT WHITMAN, "SONG OF MYSELF" (1919 [1855], 109)

Soils teem with life. However, many soil organisms are intimately involved with death. From the decomposers' point of view, they are harvesting essential nitrogen and carbon compounds from other organisms to satisfy their own growth and energy requirements. In the process, they recycle nutrients, ensuring that life can be renewed after death. Decomposition can even clean soils of diseases that were responsible for the mortality in the first place.

The cast of decomposers includes bacteria, fungi, worms, flies, beetles, mites, and even moths, whose larvae feed on hair that is difficult for most other decomposers to digest. In the Spanish language, mushrooms, the fruiting bodies of fungi, are appropriately called *carne de los muertos,* or "flesh of the dead." Predators gather at corpses to feed on the decomposers. The arrival times of parasitoid wasps and of predatory beetles and flies, and the growth rates of fly larvae after egg laying are so predictable that forensic scientists use them to estimate time of death (pl. 13). In the absence of oxygen, decomposition processes can also occur by bacterial fermentation.

Fire is another form of decomposition through very rapid oxidation. Some nutrients are carried away by heat and wind,

Plate 13. Decomposition of an elk carcass at Prairie Creek State Park.

but wildland fires have a fertilizing affect on soils overall because the ash releases nutrients that had been locked up in the standing vegetation.

Organism numbers in productive soils are virtually incomprehensible. The extremely small, single-celled bacteria are the most numerous; a teaspoon of a forest soil may hold one billion bacteria! Beneath coniferous trees, a teaspoon of soil may contain 40 miles of fungus fibers (called "hyphae"). Thousands of amoebae and other single-celled protozoans and hundreds of nematodes can occupy a teaspoon of fertile soil. But because California's Mediterranean-style climate includes winter rains and long summer droughts, much of the state's soil undergoes wide seasonal swings in organism numbers and activity. The rainy season coincides with the most rapid growth and reproduction, and the drought of summer and early autumn reduces activity. The vast acreage of desert soils in California, as well as the state's thin alpine soils, generally support fewer organisms than more productive soils do.

Larger multicellular soil organisms, such as worms, mites, insects, centipedes, and burrowing rodents, directly affect the structure of soils by digging channels and tunnels, mixing the soil, and facilitating the recycling processes of the microscopic decomposers. Without the help of shredders to break apart large particles, bacteria and fungi could not accomplish nearly so much. Mixing and burying shifts organic material into the decomposers' "work rooms" below ground. Although large enough to be seen without magnification, many of these animals remain hidden from our awareness because they are seldom seen, and so are called "cryptozoans" (derived from the Greek *kryptos,* meaning "hidden").

A classic survey was conducted in England in 1948 to estimate the total numbers of soil arthropods (just the insects, spiders, and centipedes) in the topsoil of one acre of English pastureland. Mites and springtails far outnumbered other arthropods according to the results (table 1).

TABLE 1. Total Numbers of Soil Arthropods in the Topsoil of Pastureland in One Area of England

Insect	Total Number
Mites	666,300,000
Springtails	248,375,000
Root aphids and other sucking bugs	71,850,000
Bristletails	26,775,000
Centipedes and millipedes	22,475,000
Beetles	17,825,000
Other arthropods	15,200,000

NOTE: From Salt et al. (1948).

Decomposers, Partners, and Parasites

Bacteria

Bacteria perform many tasks besides nitrogen fixation in soil ecosystems. Some species of these microscopic single-celled organisms make other essential elements available to plants, including sulfur, phosphorus, potassium, magnesium, calcium, and iron. Other species manufacture and release hormones that stimulate root growth. Bacteria protect plants by exuding a slime that can trap disease-causing microbes. They also stabilize soil structure and improve water permeability by binding soil particles together. Plants do not just passively benefit from such activities. They actively attract bacteria into the rhizosphere, the zone around roots, by secreting sugars called "exudates" that are food for bacteria.

Many bacteria manufacture chemical weapons and defenses to compete with or protect themselves from other microbes. We know those chemicals as "antibiotics." Soil bacteria were the sources of extremely useful antibiotic medicines such as penicillin, erythromycin, and tetracycline. Some bacteria can even detoxify soils. The respiration of certain anaerobic bacteria chemically transforms pollutants such as arsenic, uranium, and PCBs (polychlorinated biphenyls) into more benign chemicals, so these bacteria are used to clean up contaminated soils.

Actinomycetes are a special category of bacteria that grow long branching filaments. They are responsible for the "earthy" smell of freshly exposed soil. Many contemporary Californians may not be familiar with that wholesome odor because they live and work in urban settings that have kept dirt at a distance. Some nitrogen-fixing actinomycetes live in root nodules of nonleguminous plants, including bitterbrush, mountain mahogany, cliffrose, and Ceanothus.

Actinomycetes are especially adept at breaking down chitin and cellulose, plant tissues that are notoriously tough to decompose. They can be active at high (alkaline) pH levels where other bacteria cannot thrive. The antibiotic streptomycin is produced by actinomycetes.

Fungi

Fungi are no longer considered part of the plant kingdom, but are now classified separately as the fungus kingdom. They are saprophytes, which means that they get their nourishment by feeding on dead or decaying organic matter; in the process, they help decompose and recycle all organic matter, both plant and animal. Although fungal cells are microscopic, they often grow as long strands called "hyphae" and become visible when massed together. But our awareness of fungi begins most often after mushrooms appear above ground. Mushrooms are the fungal reproductive bodies that release spores into the air. They are the "tip of the iceberg," indicating that fungal hyphae are busy out of sight below ground (pl. 14).

Fungi degrade chitin and cellulose compounds that bacteria are unable to digest, and they can thrive in more acidic soils than bacteria can tolerate. They also reduce soil erosion because their hyphae wrap around and bind soil particles together with gluey excretions.

Almost all green plants benefit from close associations with mycorrhizal fungi (from the Greek *myco,* meaning "fungus," and *rhizal,* meaning "root"). Hyphae extend the surface area and reach of plant root systems, sometimes by hundreds or thousands of times. The strands collect water and nutrients, often more efficiently than the plant roots can, and share their bounty with host trees and shrubs. In turn, mycorrhizal fungi receive sips of sugar from the photosynthesizing plants. Hyphae are finer than root hairs and can grow into spaces too small for roots (pl. 15). The result is

Plate 14. A *Boletus edulis* mushroom, the reproductive body of a far more extensive network of fungal hyphae below the surface.

Plate 15. Mycorrhizal fungi reach outward from a plant root.

a grid of mutually beneficial connections between trees and neighboring plants that gives the concept of "plant communities" new meaning.

Fungi that surround and attach to the outside of a root are called "ectomycorrhizal" and are commonly associated with forest trees. Examples are *Boletus, Aminita,* truffles, portobellos, and the common white button mushrooms sold in grocery stores. Members of another group of fungi insert their structures partially inside plant root cells, and so are called "endomycorrhizae." They penetrate the roots but also extend long hyphae threads into the soil.

The health of all coniferous trees and of California's oak trees depends on symbiotic relationships with mycorrhizal fungi (pl. 16). In addition to aiding absorption of water and nutrients, the fungi can actively protect their host trees. They can manufacture chemicals that inhibit attacks by bacteria or other fungi, and even by browsing animals. Some fungi use looped hyphae threads to build snares that

Plate 16. A mushroom that emerges close to a tree is a sign that there may be mycorrhizal fungi underground associating with the tree roots.

snap shut to trap and kill root-eating nematode worms that come too close.

Deep tilling with farm plows and soil disturbance by bulldozers can break apart the network of fungal hyphae and lessen the fertility of those soils. When exotic plants establish themselves on a disturbed piece of ground, they can temporarily take advantage of nutrients supplied by fungi. However, because they have not evolved to give something in return, as native plants do, soil fungi can ultimately be starved out when nonnative species displace the original plant community.

One soil fungus that lives in the dry, alkaline soils of the southern San Joaquin Valley causes symptoms in humans that may be mistaken for a cold or the flu. "Valley fever" is caused by the inhalation of fungal spores. Most people get mildly sick and never suspect the true cause, but thereafter have immunity against the disease. Those with weak immune systems can subsequently develop a serious, chronic lung disease.

The balance between bacterial numbers and fungi varies according to predictable patterns among types of soil ecosystems. Most of the living matter in grasslands and farm soils is bacterial, but fungi predominate in more acid forest soils. The ratio of fungal to bacterial biomass can be 10-to-1 in oak woodlands and up to 1,000-to-1 in coniferous forests.

Biocrusts

On arid soils the bodies of cyanobacteria, green and brown algae, mosses, lichens, liverworts, fungi, and bacteria may all bind together in photosynthesizing carpets called biological soil crusts. The microbial community can crunch underfoot when you walk across such desert ground. In California biological soil crusts are found in the Mojave, Sonoran, and Great Basin deserts and on semi-arid lands of the San Joaquin Valley, Coast Ranges, and southern California coastal plains. This organic surface layer fixes nitrogen, stabilizes soil, prevents erosion, and can inhibit exotic weeds. The soil covering can also reduce water loss through evaporation. Surface "pinnacling" occurs in places where filamentous cyanobacteria and green algae swell when wet, extending upward within a crusty sheath, and then shrivel as they dry, with the sheath remaining in place. The rough texture of the pinnacles captures blowing dust and seeds, and it helps hold loose, sandy soils in place by slowing or preventing wind erosion. Domestic livestock grazing, hiking, biking, and off-road driving can damage these fragile, ecologically important crusts (pl. 17).

Arid area soils may also be capped by "desert pavement," a crust formed solely through mechanical and chemical, rather than biological, processes. Fine materials sift down below heavier gravels and, over thousands of years, form hardened, interlocking layers along the surface that inhibit plant growth, often leaving the land barren.

Plate 17. A biological crust caps sandy soil in the Mojave Desert.

Shredders, Grazers, and Predators

Protozoans

The microscopic single-celled protozoans are most common in the upper few inches of moist soils, where they can find bacteria to prey upon as well as the moisture that they require to survive. Even in soils considered to have low fertility, protozoans may number 1,000 per teaspoon of earth. In very rich soils there may be a million per teaspoon. Protozoans are classified by their method of propulsion: flagellates use whiplike flagella to move, ciliates have many hairlike cilia with which they "row" their single cell, and amoebae extend pseudopods into which they ooze forward. Protozoans can explore extremely small cavities to find their even smaller bacterial prey (pl. 18). They also eat other protozoans, algae, and sometimes fungi.

Because they have a lower concentration of nitrogen in their cells than the bacteria they consume, protozoans release the excess into the soil as ammonium, an available fertilizer for plants or bacteria. Preying on bacteria can actually stimulate the bacterial population's increase,

Plate 18. An oval amoeba is much larger than the tiny specks of bacteria nearby, but slightly smaller than the angular sand particle. Protozoans feed on bacteria, helping to cycle nutrients in the soil.

a response that is analogous to the stimulated growth that follows the judicious pruning of fruit trees. Protozoans themselves serve as food for larger predators, such as nematodes.

Mites

There can be 100,000 to 400,000 mites in each square yard of moist forest soil. They are not insects but are members of class Arachnida, which also includes spiders. Some are predatory, but most are scavengers that eat fungi and leaf litter in the top organic layer of soils, shredding it into small pieces that can then be worked upon by microbial decomposers. Mites have a protective shell that can be opened and closed, making them look like dark little seeds or tiny beetles (pl. 19).

Plate 19. Mites are microscopic, and they occur in incredible numbers in the soil. This image of a mite on fungus is magnified 850 times.

Springtails

California has about 150 species of springtails, which are normally less than a quarter inch long and hard to spot, but can become easily visible when they swarm by the millions out of the ground, showing up as flecks of white springing around a pile of dirt or leaves. They are usually less abundant than mites, although the two types of arthropods often share the same ground. Springtails have six legs, like insects, and used to be classified as insects, in the order Collembola, but they are now considered to be their own class. They eat fungi primarily but also live on plant material, and a few species prey on small animals such as nematodes. Photographs have caught them ingesting roundworms, starting at one end as if sucking up a strand of spaghetti. Their waste enriches soils with nutrients that become available to plants.

The scientific name Collembola comes from Greek words meaning "glue-peg." Most springtails have tubes on the lower sides of the abdomens that absorb water but can also exude a sticky substance that helps them attach to surfaces. The common name "springtail" refers to an appendage at the tail end that is normally tucked forward underneath the body. It can snap suddenly down to launch them into the air, allowing a springing escape from attacking predators such as ants. Species that live entirely underground have no room for jumping and do not have the springing appendage of their relatives (or the bright coloration, another useless characteristic in the dark underground).

During winter thaws, when springtails sometimes mass against snowbanks, people call them "snow fleas." No one is certain why they swarm, but population growth pressures may trigger the events.

Springtails, like a number of other primitive soil animals, exhibit some intriguing reproduction adaptations. Males have no external sexual organs with which to copulate

Plate 20. This spermatophore (a blob of semen on a stalk used for indirect fertilization) was produced by a springtail in the family Isotomidae. The springtail here is of another species, *Kalaphorura burmeisteri,* and may intend to eat the spermatophore, which is a rich source of protein.

with females. Instead, in a number of springtail species, males attach vertical stalks to the ground by exuding a liquid that hardens on contact with air. Then they deposit a blob of semen on top of each stalk (pl. 20). The males will sometimes build stalks in a circle around females and then maneuver females into position to sit on sperm packets and become fertilized. In other species the males build spermatophores in a cluster that looks to the unassisted human eye like a patch of fuzz. At least one fascinated writer has referred to these patches as "love gardens," because sperm is transferred to female springtails as they amble through and brush against the stalks. Males also patrol these sites regularly to replace aging sperm packets with fresh ones. Some mites, centipedes, and millipedes also lack external sex organs and must accomplish reproduction similarly through indirect fertilization, with males depositing sperm packets for females to encounter.

Nematodes

Minuscule and eel-like, nematodes have no segments (unlike earthworms) and are commonly called "roundworms." Typically, nematodes are only one-twentieth of an inch (about a millimeter) in length. Although too small to attract our notice, they are incredibly abundant in nearly every soil, residing in the water films surrounding soil particles and sometimes numbering hundreds of millions per acre (pl. 21). They also live on the ocean floor and on lake bottoms, and as parasites inside plant and animal hosts. There are tens of thousands of species. Their omnipresence led N. A. Cobb to write (in the *1914 Yearbook of the United States Department of Agriculture*): "If all the matter in the universe except

Plate 21.
A root-knot nematode larva, *Meloidogyne incognita,* entering the root of a tomato plant (magnified 500 times).

the nematodes were swept away, our world would still be dimly recognizable, and if, as disembodied spirits, we could then investigate it, we should find its mountains, hills, vales, rivers, lakes and oceans represented by a thin film of nematodes" (1915, 472). That fanciful image could work as well if applied to bacteria or fungi, the other nearly ubiquitous soil species.

Nematodes fill the whole range of food-gathering niches: some species eat plants, others feed on bacteria and fungi, and still others prey on protozoans or other nematodes; some are omnivorous and select from the entire menu. Beneficial species enhance soil productivity by mineralizing nutrients (passing them through the gut into the soil) or by preying on disease-causing organisms. Like protozoans, nematodes release ammonium after feeding on bacteria and fungi, food sources that contain more concentrated nitrogen than the worm requires. Nematodes provide a benefit to bacteria, which cannot move on their own, by transporting them through the soil in their digestive systems, or as riders stuck to the outside of the worms' bodies.

Predatory nematodes can be purchased for use in gardens and nurseries as a form of organic pest control. Others, such as root-knot nematodes (*Meloidogyne* spp.), are parasites that take residence inside plant roots and damage crops. Galls form on the roots, and the feeding roundworms sap the plant's vigor, causing wilting and leaf yellowing. Management of harmful nematodes in agriculture and gardens focuses on the use of resistant varieties of plants and on preventing introduction of the roundworms at the time of planting. Agricultural inspection stations at the state's borders aim at blocking the entry of "bad" nematodes into California by keeping out plant material that has soil on its roots.

Certain marigold (*Tagetes*) species suppress some types of soil nematodes when planted in place of a crop for an entire season. Chemical pesticides and fumigants are not generally feasible methods of controlling nematodes because they harm so many other, essential soil organisms.

A few nematode species are intestinal parasites for animals. Larvae of hookworms, for example, can enter a host dog, cat, or human whose skin comes in contact with contaminated soil. The worms mature in the host's intestines, and eggs are passed back to the soil in feces.

Earthworms

The stars of the underground show are earthworms, because of their role in building soil, their importance to gardeners, and the fascination they afford schoolchildren. Earthworms are giants compared with most other soil organisms. The earthworm's body is made of segmented cylinders, with a mouth at the head and an anus at the tail end. They are omnivores that chew their way through soil, feeding on plant and animal detritus and on microbes, digesting and thoroughly mixing organic and inorganic components before eliminating it as humus. This process breaks complex organic molecules down into simpler forms more suitable for uptake by plants. When plant material is scarce below ground, earthworms come to the surface to find leaves, which they pull into underground chambers (pl. 22).

Charles Darwin was fascinated by earthworms and conducted many experiments on them to get a greater understanding of soil formation. He estimated that English earthworms manufactured an inch of topsoil every century. He wrote, "The plough is one of the most ancient and most valuable of man's inventions, but long before he existed the land was in fact regularly ploughed, and still continues to be thus ploughed by earth-worms. It may be doubted whether there are many other animals which have played so important a part in the

Plate 22. Nightcrawlers are large earthworms that pull food from the surface down into vertical burrows.

history of the world, as have these lowly organized creatures" (Darwin, 1881, 313).

Material that has passed through the gut of an earthworm is of a size just right for redigestion and further decomposition by progressively smaller microorganisms. Earthworm feces, called "casts" or humus, have higher concentrations of carbon, nitrogen, phosphorus, and other nutrients compared with the surrounding soil. The burrowing worms also increase the number of soil pores, improving water infiltration and the movement of air into the ground. Roots take advantage of worm channels as paths of least resistance for growth. Roots also benefit from the coating of nutrients that the worms deposit along the walls of their tunnels.

There are three categories of earthworms, based on where they reside in the soil. Litter-dwelling, epigeic worms live in the uppermost layers of soil. Shallow soil-dwellers, the endogeic worms tunnel horizontally one to two feet below the surface. Deep burrowers, called anecic worms, live in permanent vertical burrows as much as six feet down, although they come to the surface to forage (fig. 6).

Groups of Earthworms

1. Soil surface-dwelling or compost preferring species
2. Topsoil-dwelling species
3. Subsoil-dwelling species

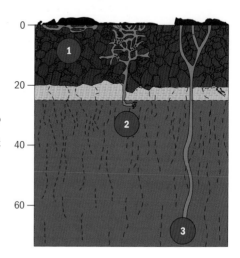

Figure 6. Depths where earthworms reside. Earthworms can basically be divided into three groups, with each group preferring to live in a distinct area within the soil.

Lumbricus terrestris, the familiar Nightcrawler used so often as fish bait and common in suburban gardens, is of the anecic type. The species was introduced to California from Europe. Such deep dwellers feed at the surface only at night because long exposure to ultraviolet radiation in sunlight is lethal. Nightcrawlers are not a particularly good worm choice for a composting bin, because they prefer to shelter deep down in undisturbed burrows. A better choice for garden composting is another introduced species, the Red Wriggler *(Eisenia fetida),* which is smaller and redder and does not have the dark-colored head of the Nightcrawler.

Most native Californian earthworm species are the endogeic type, generally occupying horizontal tunnels at middle depths. Their activity peaks during the rainy season, but they tolerate drier soils. Virtually no native worms are found where plows

VERMICULTURE

Vermiculture is the practice of growing worms to produce garden compost. Red Wigglers are commonly used for composting. They have a big appetite, reproduce quickly, and can thrive in containers. They can eat more than their own weight in food every day and recycle much of a home's food waste, including fruit, vegetable, and bread scraps. My son raised worms in his college dorm room and was amazed when a large piece of moldy pumpkin disappeared in a single day. Vermiform bins do not smell bad so long as the number of worms remains in balance with their food and the contents are stirred to keep them aerated. Meats, dairy products, oily foods, and grains should not be put into the bins. Shredded newspaper or cardboard is often used as bedding, but fallen leaves, straw, or sawdust also will work. The brown, crumbly worm casts can be periodically harvested and used as mulch for plants and to enrich garden soil.

or bulldozers have churned the soil, where trees and shrubs have been cut down, on irrigated farmland, or in typical suburban housing tracts. Worms found in such places are almost always exotic species.

Not much is known about native California earthworms. Valuable information was lost when an extensive collection was destroyed in the 1906 San Francisco earthquake. Some native California worms are in the family Megascolecidae. Their translucence distinguishes them from the more commonly seen, opaque Nightcrawlers. Other visible, but more subtle, differences include sex characteristics and the positioning of structures on segments (fig. 7).

In 1993 a survey of native earthworms was conducted on southern California wildlands from Santa Barbara County

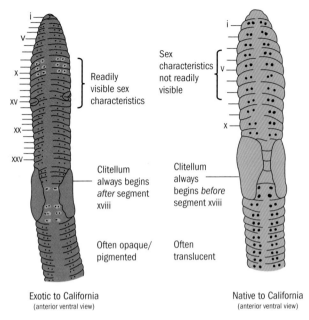

Figure 7. Comparison of exotic (Lumbricidae) and native (Megascolecidae) earthworms in California.

to San Diego County. Native species preferred grass and oak habitats but were also found in chaparral. They were rare or absent in the more acidic soils of coniferous tree stands. They tolerated drier soils and harsher conditions than did exotic earthworms, and they apparently out-competed exotic species where the land remained undisturbed. Few Californians are aware of them, however, because they spend so little time at the surface.

Shrews and moles feed on earthworms, which are also food for birds. Robins, for example, listen for worms moving just below the surface and plunge their beaks into the ground to snag them. Earthworms, like everything else that dies, eventually surrender their bodies to the microbial decomposition and recycling crews.

In farm ecosystems, practices that promote the growth of earthworm populations include the use of organic amendments such as composts and cover crops (green manures) and reducing tillage (which also reduces compaction). Plows can destroy vertical worm burrows and can cut up the worms themselves. Tillage also dries soils and can bury material that surface feeders depend upon.

As for that intriguing behavior of worms appearing on the surface after heavy rain, the belief I shared with most of my friends during childhood—that they were escaping from drowning—is probably not accurate. Earthworms actually can survive quite a while in oxygenated water, because they absorb the gas directly through their skin. They may, in fact, be seeking mates. Researchers have noted that most of the worms on the surface at such times are large and sexually mature, although the need for a breath of dry air would force up worms in every age group. Perhaps the surface moisture improves the conditions for successful reproduction.

Earthworms are hermaphrodites: both male and female sex organs are found on each worm. During reproduction, earthworms attach to one another, and sperm is discharged into both individuals' female openings simultaneously. Fertilized eggs, resembling small grains of rice, are shed into the soil.

Ants

In arid and semi-arid soils, ants are the most numerous soil arthropods, and there are far fewer springtails and mites. Such lands are too dry for worms, but ants (and termites) carry out the tasks of "ecological engineering": tilling and redistributing soil matter, aerating, decomposing matter, and piling soil. "If ants went on strike and ceased their ecological services, the consequences would be profoundly disruptive to the natural world—and eventually tragic for humanity" (Fisher and Cover, 2007, xii). Workers in Harvester Ant (*Pogonomyrmex californicus*) colonies in California desert

Plate 23. Red ants are the "environmental engineers" of the desert.

areas can create galleries 15 feet deep and move hundreds of pounds of subsoil to the surface as they excavate tunnels and galleries. Near their holes they deposit "compost heaps" of debris from the foods they have scavenged. Ants add to the seed crop in the soil near their nests by moving millions of seeds to their granaries below ground (pl. 23).

Some ants tend aphids to "milk" them for honeydew nectar. Red mound ants of the genus Formica make mounded nests of pine needles in coniferous forests and will swarm out of the ground to bite anyone who lingers too close for the ants' comfort.

The very large, shiny black ants that tunnel into fallen logs or stumps, called Giant Carpenter Ants (*Camponotus laevigatus*), can extend galleries into the ground and sometimes bore into wooden structures built by humans. Despite their impressive size, they seldom attack people.

Ants could hold important clues to improving water clarity at Lake Tahoe and to maintaining ecological health in the Lake Tahoe basin. "Aerator ants," one of the more common categories at Tahoe, construct extensive tunnel networks in the ground. Those channels facilitate water infiltration in forest soils and decrease surface runoff, which can carry sediments into the lake. In forests subject to urban land development, the abundance of aerator and decomposer ants has dropped precipitously. Researchers have suggested that these ant communities and their ecosystem services be considered by land use planners at Lake Tahoe.

Termites

Nature's wood recycler, the termite, relies on symbiotic bacteria and protozoans that live in its gut and digest cellulose. Termites construct large colony nests above or below ground that may house thousands of individuals. There are several species of termites in California, including the Western Subterranean Termite *(Reticulitermes hesperus)*, the most common termite in the state and the most destructive to human-built wood structures. Termites process much of the leaf litter in the desert and can be a keystone species there, benefiting the ants, springtails, mites, and beetles that share their extensive burrows (pl. 24).

Plate 24. Western Subterranean Termites are native to California.

Pill Bugs

Touch a pill bug and it may roll up to protect itself, an act called "conglobulating." If a pill bug conglobulates in California, it is *Armadillididae vulgare,* and not a native species (pl. 25). Common sow bugs *(Porcellio scaber)* are also exotic in California, although native sow bugs, or wood lice *(Ligidium gracile),* dwell in California forests. All are crustaceans with segmented outer shells and are also known as isopods (*iso* meaning "equal" and *pod* meaning "foot"): they have an equal number of similar legs on each side. They are omnivorous scavengers that eat plant detritus and so aid the decomposition process occurring in soils. Pill bugs also roll themselves up to conserve moisture. Often they are seen only when logs, rocks, or garden flowerpots are turned over.

Plate 25. Pill bugs that roll up when threatened are not native to California. Their shredding of plant material provides fodder for smaller soil organisms.

Plate 26. This Yellow-spotted Millipede *(Harpaphe haydeniana)* does not actually have a thousand legs, as its name suggests, but rather two legs on each body segment. Bright color spots warn potential predators that the millipede can fight back with a chemical secretion of almond-scented hydrogen cyanide.

Millipedes

The many-legged millipedes, of class Diplopoda, look impressively scary but are herbivores, feeding on leaf litter and fungi. As a result, they influence decomposition by shredding plant detritus, and their droppings contribute nutrients to humus (pl. 26).

Dung Beetles

Wherever animals graze, the dung beetle, a type of scarab beetle, may be found excavating burrows under piles of dung. Dung beetles take advantage of the nutrition in cattle feces by laying eggs in a portion of the dung, which they roll into a ball and bury. Cows and other grazing animals generate large amounts of dung. The recycling of those wastes is accelerated by the lowly dung beetle, and the decomposition also helps limit the numbers of livestock intestinal parasites and

Plate 27. The native Morro Shoulderband Snail *(Helminthoglypta walkeriana)* is an endangered species. This one was photographed at Montaña de Oro State Park.

flies. California has a variety of species, including the Spotted Dung Beetle *(Sphaeridium scarabaeoides)* and an introduced European species, *Aphodius fimetarius.*

Snails and Slugs

Gliding along on muscular "feet," snails and slugs *(Gastropoda* spp.) constantly secrete mucus slime trails; such trails, once dry, may be the most obvious evidence of their activity. Slugs lack the external shell that snails carry. Snails and slugs hibernate in the topsoil through cold periods. During hot, dry spells, or when it is very cold, snails attach themselves to something solid and use a membrane to seal the edges of the shell against the outside air.

California has 279 snail species, 37 of which are exotics. The Brown Garden Snail *(Helix aspersa)* is a common garden pest, introduced from France in the 1850s by escargot fanciers. There are over 60 species in the group of native California

Plate 28. On the day this banana slug was photographed, it was the only one seen crawling around. Several others were seen in holes along the same mossy hillside, waiting for wetter conditions before venturing out.

land snails called "Shoulderbands" (*Helminthoglypta* spp.). They look like garden snails but have a distinct, dark band that wraps around the "shoulder" of the shell. Included in this group is the Morro Shoulderband Snail *(H. walkeriana),* an endangered species endemic to San Luis Obispo County, with critical habitat in Montaña de Oro State Park (pl. 27). Native snails are found primarily in less disturbed native ecosystems and are unlikely to be seen in gardens.

In California's moist central and north coast forests, a number of banana slug species (*Ariolimax* spp.) participate in the shredding and recycling of forest litter at the surface. The Pacific Banana Slug *(A. columbianus)* is found in the northern and central Coast Ranges redwood forests (pl. 28). Smaller populations are found in the northern Sierra Nevada as far south as Tuolumne County, and in the Palomar Mountains of southern California. With slimy bodies up to eight inches long, they are yellow (sometimes greenish or tan) and may have dark blotches. The Slender Banana Slug *(A. dolicophal-lus)* is the athletic mascot of the University of California at

Plate 29. Scorpions are key predators in desert soil ecosystems.

Santa Cruz. It is usually light yellow, with no blotches, and looks very similar to the central coast's *A. californicus*.

Higher-Level Predators

Antlions

Across most of California, where there is loose sand, funnel-shaped pit traps can be seen. They were built by antlion (*Brachynemurus* spp.) larvae, which then buried themselves at the base of each cone. Insects that stumble into the funnel try to scramble up the steep walls but often fall to the bottom, where the antlion's jaws await. These predator-prey dramas go unnoticed by humans—most of the time. The antlion inspired a monster in the film *The Return of the Jedi*, in which, on the planet Tatooine, a massive Sarlacc waited at the bottom of its sand pit for Luke Skywalker.

Plate 30. Tarantula burrows are lined with web silk.

Scorpions

The dominant predators in desert soil ecosystems might be scorpions, which are arthropods in the order Scorpiones. Their flattened bodies allow them to hide under rocks and logs until they emerge at night to hunt (pl. 29).

Centipedes

With fewer legs than millipedes, the otherwise similar-looking centipedes (class Chilopoda) are meat eaters that prey on insects and other small soil-dwelling animals. Because their eggs develop in underground nests, female centipedes must protect the eggs from soil fungi by licking them, which applies a fungicide in their saliva.

Tarantulas and Trapdoor Spiders

Very large spiders, called tarantulas (*Aphonopelma* spp.), are found across much of California. Those commonly seen

Plate 31. A female trapdoor spider closing her door after it was opened by the photographer.

wandering during September and October are likely males in search of mates. Tarantulas are present all year but become much more reclusive during summer, spring, and winter, emerging from their underground burrows only at night to hunt for food. Their small burrows are lined with silken web material (pl. 30).

The females of another group in the tarantula family, trapdoor spiders (*Antrodiaetidae* spp.), construct tubes lined with web silk just wide enough for the spider, with one wider place farther down in the tube where she can turn around. The surface opening is tightly sealed with a door made of silk and earth, camouflaged with material stuck to it, and weighted so that the door falls closed on its own. The spider will hold it shut from the inside with her fangs if a predator, such as a Tarantula Hawk Wasp, comes "knocking" at the door. At night trapdoor spiders wait at the open door, watching for prey to come within reach (pl. 31).

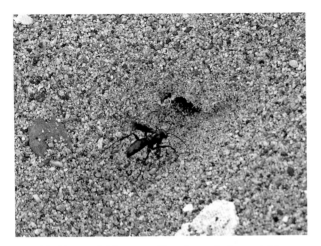

Plate 32. After she finishes excavating a burrow in the sand, the female spider wasp will bring a paralyzed spider into the burrow to provide food for the larvae that hatch from the eggs she lays in the spider's body.

Spider Wasps

More than 100 species of small black or steel-blue wasps that specialize in preying on spiders (family Pompilidae) are found in California. On a sandy beach, such as along the shore of Mono Lake, such small, dark wasps may be seen energetically burrowing into the beach, kicking sand out of the narrow openings as they excavate. Once satisfied that she has built a proper cavern, a female brings a paralyzed spider into the hole and lays eggs in its body. When the eggs hatch, the spider becomes the young wasps' first meal (pl. 32).

Larger black wasps with orange wings in this group specialize in hunting tarantulas. They are commonly called Tarantula Hawks (*Pepsis* spp.).

Plate 33. Yellowjacket wasps love our picnic food but feed mostly on other insects.

Yellowjacket Wasps

An all-too-common pest at picnics and around garbage cans at campgrounds, black-and-yellow-banded Yellowjackets (*Vespula* spp.) are not much larger than a housefly (pl. 33). They are "social wasps" that nest in colonies below ground or in hollow logs or other cavities. Despite their troublesome interest in sodas and hot dogs, the wasps have a beneficial role as voracious predators of other insects, including many flies, caterpillars, and beetles that are home, garden, and farm pests. They also scavenge from decomposing bodies and rotting plant material. They in turn become food for birds, skunks, raccoons, and bears.

In California, the Western Yellowjacket (*Vespula pensylvanica*) is a widespread native species. A queen establishes a nest site, often in an abandoned rodent burrow, and constructs a tan, papery nest made of masticated wood pulp and containing small chambers, where she lays eggs. Larvae emerge as sterile females that will feed later larvae batches. New generations keep expanding the nest and excavating to enlarge the hole. Their numbers peak at the end of summer, which coincides with the hatching of fertile females and males. After mating, the males die. The old queen and workers also die when winter comes, but newly fertilized queens spread out to find shelter through the winter and establish new colonies in the spring.

Yellowjackets love the meat protein and high-energy carbohydrates in picnic foods, which they bring back to the nest to feed the growing larvae. Unlike honeybees, which have barbed stingers and die after stinging once, wasps have smooth stingers and can sting repeatedly.

The German Yellowjacket *(V. germanica)* has also been found in California since the early 1990s, when it moved into the state from the Pacific Northwest. This wasp is more likely to nest above ground in tree cavities or in structural walls.

Burrowers for Shelter and Food

Some animals excavate burrows that they use for shelter part of every day, or during periods of hibernation. Other species spend very long portions of their lives underground, emerging for relatively brief intervals to feed and reproduce.

Spadefoot Toads and Cicadas

Adult Spadefoot Toads (*Scaphiopus* spp.) appear above ground only during wet seasons to eat and breed. Their eggs and larvae develop quickly, in just a few weeks, while there is still water in breeding pools. The mature adults burrow back into the mud using "spades" on their hind feet as digging tools, and then wait, their metabolism slowed, until the drumming of raindrops sends the message that it is time to come up for activity again. The Couch's Spadefoot Toad *(S. couchii)* lives in the desert lands of southeastern

Plate 34. A Couch's Spadefoot Toad, up from the mud.

California, in scattered populations east of the Algodones sand dunes of Imperial County, and north into San Bernardino County (pl. 34). Western Spadefoot Toads *(S. hammondii)* are endemic, found only in a range that extends from the north end of California's great Central Valley, near Redding, down to northwestern Baja California. That takes in a variety of habitats, including mixed woodlands, grasslands, chaparral, sandy washes, river floodplains, alluvial fans, playas, alkali flats, foothills, and mountains. The Great Basin Spadefoot Toad *(S. intermontana)* may spend eight months buried underground during the cold winters and dry summer periods that characterize the Great Basin high-elevation desert. They breed in the late spring or summer in temporary pools, but also along lakeshores, in irrigation ditches, and even in livestock water tanks. In California this species is found east of the Sierra Nevada from the Oregon border down to the northern Owens Valley.

The cicada is another animal that spends a long time underground. California has 65 species of these insects, whose eggs mature on shrub or tree branches. Nymphs then drop to the

Plate 35. Cicada nymphs may spend years underground before emerging as adults, intent on reproduction.

ground and burrow down, feeding on plant roots and fungi, not emerging as adults until years later. It is only in eastern North America that cicadas have the extremely long, 17-year emergence cycle. Cicadas in California may reappear after two to five years. Males make the conspicuous noises that draw our attention, although the individuals creating the racket can be hard to spot. The male hopes to attract females by vibrating a pair of drumlike abdominal membranes. Particularly noisy outbursts are produced by the Great Basin Cicada *(Okanagana cruentifera)* (pl. 35). In the Colorado Desert, Apache Cicadas *(Diceroprocta apache)* call loudly from perches in Palo Verde trees. The clicking noise made by the smaller Woodland Cicadas (*Platypedia* spp.) in the Sierra Nevada foothills and mountains are manufactured using their wings.

Desert Tortoise

With legs adapted for burrowing, Desert Tortoises *(Gopherus agassizii)* excavate underground homes that are sometimes more than nine feet long, places to escape the intense desert

Plate 36. The Desert Tortoise has powerful front legs designed for digging.

heat as well as to hibernate during the cold Mojave Desert winters (pl. 36). They also gouge catch-basins in the soil, to which they come back for drinking water when it rains. The Desert Tortoise is listed as a threatened species in the Mojave and Sonoran deserts under both state and federal law. It is unlawful to touch, harass, or collect these animals.

Moles and Gophers

Both moles and gophers till the soil, helping mix nutrients within the root zone. Their tunnels also provide habitat for other species. The California (or Broad-footed) Mole (*Scapanus latimanus*) spends its life underground, so this rodent is very rarely seen by people. Young moles dispersing from where they were reared might be seen traveling above ground. Otherwise, the only signs are conical molehills or arched ridges of earth disturbing the surface where the moles have tunneled along, using their heavily clawed front feet as shovels and occasionally ejecting earth at the surface. They move below ground to hunt, seeking

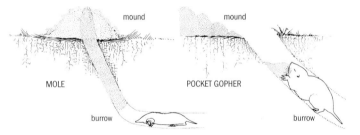

Figure 8. Moles push the soil they excavate up through a mound, whereas gophers push the dirt outward in a fan pattern from a hole at the surface.

encounters with their food: earthworms, beetles, and insect larvae. Moles are considered insectivores, although they occasionally feed on seeds or roots. They are extremely sensitive to vibrations transmitted through the soil, which helps them locate prey. Because they have little use for eyesight, they lack the ability to see, other than distinguishing light from darkness. They dig deeper tunnels to permanent burrows used for resting and nesting. The activity of moles and gophers can be distinguished because moles make conical mounds of dirt at the surface to clear their tunnels, but gophers typically expel dirt in an outward fan around the opening (fig. 8).

Pocket gophers (*Thomomys* spp.) spend very little time on the surface, usually emerging only at night, when they may feed on surface plants. Below ground, they seek out roots to eat and may pull the above-ground portion of a plant, leaves and all, down into their tunnel (much to the dismay of gardeners). Material may end up in a storage area in their multi-room dens. Gophers are dug from their burrows and eaten by coyotes, badgers, and skunks and can be attacked underground by snakes and weasels. California is home to the Valley Pocket Gopher *(T. bottae)* and the Sierra Pocket Gopher *(T. monticola).* As winter ends and snow melts, "gopher cores" sometimes appear where

Plate 37. Gopher runs are visible evidence that life is being carried on underground.

gophers had packed soil along tunnels just beneath the snow pack (pl. 37).

Ground Squirrels

The California Ground Squirrel *(Spermophilus beecheyi)* constructs extensive tunnel systems with specialized rooms for sleeping and storage and with several openings to the surface that may be used by other members of a squirrel colony (pl. 38). They are found across all of California except the deserts, although they have been extending their range into the Great Basin uplands east of the Sierra Nevada in recent decades. They are omnivores, as are the Belding's Ground Squirrels *(S. beldingi)*, which live in the Sierra Nevada and the adjacent Great Basin uplands.

Plate 38. California ground squirrels excavate extensive underground chambers with multiple openings to the surface.

Belding's are sometimes called "picket pin" ground squirrels because of their habit of standing up straight and tall to peer about. Sharp whistles alert the rest of the colony when a squirrel spots something alarming. Both species typically remain underground from mid-summer until early spring to avoid the dry heat of summer and the cold of winter, emerging only when California's winter rains bring new plant growth.

Badgers

The digging machine that is the American Badger (*Taxidea taxus*) is endowed with powerful shoulder muscles and heavy foreclaws that enable it to "swim" rapidly down into the soil. Badgers are rodent-control predators, seeking mice, gophers, and ground squirrels in the tunnels of their prey. They excavate and reside in their own large burrows. The range for badgers includes the entire state of California, but they prefer undisturbed wildlands and are seldom seen (pl. 39).

Plate 39. The badger is a digging machine, looking for prey below the surface.

Bank Swallows

Holes in sandy streambanks and coastal bluffs give refuge to colonies of Bank Swallows *(Riparia riparia)*. These tiny birds catch insects on the fly in their wide-open mouths. The male locates a starter hole in a cliff and begins additional excavating, pecking at the dirt until he has attracted the attention of a female; together the pair complete the home building task. An entrance opens into a tunnel, sometimes several feet long, that is lined with grass and that slopes upward to keep the nest area dry. Colonies share the bank locations and in the winter migrate as flocks to Central and South America. Bank Swallows used to be common in southern and central California but are now listed as a threatened species as the result of habitat disruptions. They no longer breed in southern California. About 70 percent of the remaining population is found today along the banks of the Sacramento and Feather rivers (pl. 40).

Plate 40. Bank Swallows at Fort Funston, San Francisco.

Burrowing Owls

Another bird species that uses the ground for shelter is the
Burrowing Owl *(Athene cunicularia)*. These are one of the
smallest owls that exist, about the size of a quail. They have
long legs that let them see over vegetation as they stand near
their burrows. Burrowing Owls do not do their own digging, but seek abandoned ground squirrel or rabbit burrows, where they take shelter and build nests. They prey
on grassland and desert insects, small rodents, reptiles, and
small birds. The owls have been known to decorate their
burrow entrances with livestock dung, perhaps to con-
ceal their own odors from badgers or other predators. At
one time, Burrowing Owls were one of the most common
birds across California, nearly ubiquitous in coastal grasslands. Populations have declined because so much of their
favored habitat has been paved over and, perhaps, because
ground squirrels, their favored food and burrow supplier,
have been controlled as pests in many locations. Although
they are not listed as endangered or rare, they are considered

Plate 41. A Burrowing Owl in the Imperial Valley.

a "California species of special concern." Most Burrowing Owls in California today are found in the San Joaquin and Imperial valleys (pl. 41).

Not all of these organisms actively build the soil structure, but each plays a role in the soil ecosystem, occupying a trophic (literally, "feeding") level in the "who eats what" hierarchy of production and predation. The summing of these many parts produces an incredibly complex result: a truly living soil.

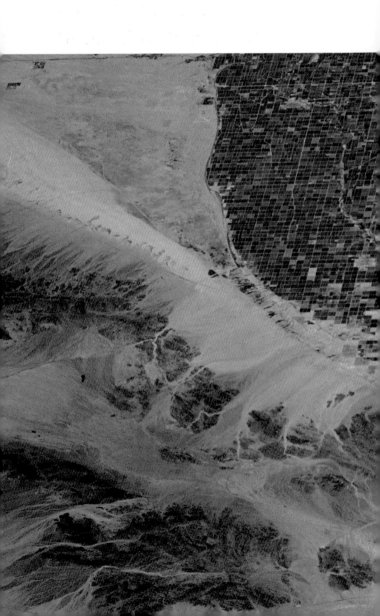

When acorn gathering, there was a family area that you go to year after year. . . . You didn't really own the trees but you knew what the trees were going to give you.

GLADYS MCKINNEY, MONO INDIAN (ANDERSON 2005, 134)

Preparing a building site meant carting the topsoil off to a landfill. Sometimes the fine topsoil was sold as fill for use in other projects. Completely paved, Silicon Valley won't feed anyone again for the foreseeable future.

DAVID R. MONTGOMERY (2007, 171)

WITHIN THE BOUNDARIES of what is now the state of California, Indians once lived in the highest population density found in North America. From 80 to 100 languages were spoken by approximately 300,000 California Indians. The native Californians were "home bodies," specializing in detailed knowledge of their local homelands. Anthropologist Theodora Kroeber saw the California Indians as "true provincials" whose defining trait was a preference for "a small world intimately and minutely known"(Kroeber, 1976 [1961], 23). Land ownership was not a feature of those cultures, although a group might claim harvest rights at specific oak trees or fishing rights at a particular river location (pl. 42).

The successors to these first Californians—Spanish mission builders and rancheros, followed by gold-seeking '49ers and settlers—found a landscape and diversity of soils that met their varied expectations and needs. Many of California's cities began as trade center villages serving the surrounding

Plate 42. A dam built by Hupa Indians to catch salmon.

farm and ranch lands. Urban development spread outward from such centers, inevitably covering some of the most productive farm soils in the state's alluvial plains, coastal valleys, and upland terraces.

The concept of owning land, of holding title to property, arrived with the Spanish and, later, Mexican governments, which distributed massive land grants to a relatively small number of rancheros. That transition began in 1769, when a Spanish expedition led by Don Gaspar de Portola came north from Mexico. The group's diarist, Father Juan Crespi, described one "spacious valley, well grown with cottonwoods and alders, among which ran a beautiful river from the north-northwest. . . . We halted not very far from the river, which we name *Porciuncula*. . . . The plain where the river runs is very extensive. It has good land for planting all kinds of grain and seeds, and is the most suitable site of all that we have seen for a mission" (Bolton, 1927, 146, 147). The San Gabriel Mission was established nearby. It was one of 21 missions the Spanish would complete by 1823 between San Diego and Sonoma, none of them very far from the coast. In 1781 a farming pueblo was founded in the vicinity of the San Gabriel mission with the ungainly title *El Pueblo de Nuestra Senora la Reina de Los Angeles de Porciuncula*. The pueblo would become known to the world more simply as "Los Angeles."

Spain made land grants to a few dozen individuals, and several hundred additional grants followed after the Mexican government took control of Alta California in 1872 (map 4). Grazing cattle and horses on the large ranchos transformed the vegetation of the coastal valleys and mountains, inadvertently introducing annual grasses from Europe that replaced many native plant species.

As one example, Bernardo Yorba's holdings included Rancho Santiago de Santa Ana (62,516 acres), Rancho Canon de Santa Ana (4,449 acres), and Rancho Lomas de Santiago (47,226 acres). "Yorba . . . in the day of his glory, . . . might have

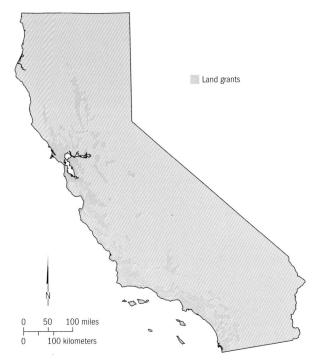

Map 4. The extent of Spanish and Mexican land grants at the attainment of statehood in California.

traveled fifty to sixty miles in a straight line, touching none but his own possessions. His ranches . . . were delightfully located where now stand such places as Anaheim, Orange, Santa Ana, Westminster, Garden Grove and other towns in Orange County" (Newmark, 1916, 177).

The land grants gradually changed hands after California became a state, in 1850. Overgrazing, drought, and new tax laws forced that transition, and federal homestead and other land laws encouraged settlement and development. Yet it was the quest for gold that would become the driving force behind the relationship between Californians and the land.

Lines on the Land

Under the February 2, 1848, treaty of Guadalupe Hidalgo, which ended the Mexican-American war, the boundary line between the United States and Mexico was to run from the confluence of the Colorado and Gila rivers westward to the Pacific Ocean, at a point one maritime league south of the port of San Diego.

The northern boundary of the new state of California was set at latitude 42 degrees north, because a treaty in 1819 had established that line, extending from the Arkansas River west to the Pacific Ocean, as the border between lands owned by Spain and the United States (later, the northern boundaries of Nevada and Utah would also follow the 42 degree line).

The eastern boundary of California became a major point of debate during the Constitutional Convention of 1849. There was no question that it had to be far enough east to include the mineral wealth of the Mother Lode soils in the Sierra Nevada. Some argued for extending the state east across the Great Basin all the way to the Rocky Mountains. Eventually the northern anchor for the eastern boundary was set where the 120 degree longitude line intersected the northern border. The line then ran straight south until it reached the thirty-ninth parallel. Those at the Constitutional Convention presumed that this would place the boundary east of Lake Tahoe (known then as Lake Bigler). Maps were drawn showing the lake entirely inside California's borders. Later surveys revealed that the corner where the oblique line began actually lay in the middle of Lake Tahoe. From that point the boundary line angled southeast, roughly paralleling the Sierra Nevada mountain crest, until reaching latitude 35 degrees at the Colorado River. From there the boundary followed the middle of the river channel down to the Mexican border (map 5).

On the ground, delineating the state's eastern boundary presented difficulties. Determining longitude was a major

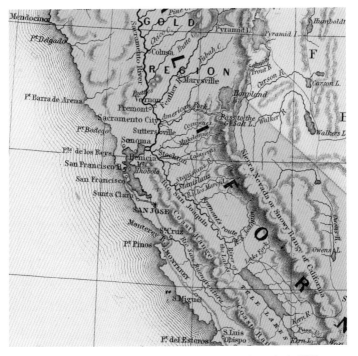

Map 5. Detail from a map of California that was rushed into print in 1850. It depicts Lake Tahoe as lying inside the state's boundaries. Later surveys would clarify that the eastern boundary actually runs into and makes a bend within the lake.

challenge for surveyors in the nineteenth century, who depended on astronomical observations. Most of the eastern boundary still had not been surveyed when Nevada became a territory in 1861. So, on February 13, a team of 14 men, led by Dr. J.R.N. Owen, left the Colorado River (near Needles) on a northwesterly reconnaissance of the boundary line. The group traveled with three horses, 22 mules, and three camels purchased from the U.S. Army (which had been experimenting with camels for desert patrols since 1855).

Consider the uncertainties facing that survey party, attempting to follow an invisible line across the desert landscape. Each day guides were sent ahead to search for water. Only water hole locations along the first part of the route were known. The group had to find the rest by reading the land and hoping that they would encounter a water source every few days.

On March 3 the party came upon a faint trail made in 1849. Twelve years earlier emigrant wagons had etched those first, persistent wagon wheel tracks. Some of those pioneers and their animals died in the arid basin to the west that, ever since, has been called Death Valley.

The reconnaissance party, unable to find enough water along the boundary route, were finally forced to leave the survey line and seek springs they knew of in Death Valley. They abandoned the northward journey on March 13 and eventually passed through the Owens Valley and around the southern Sierra to return to civilization in the Central Valley.

In 1863 another survey was commissioned by Governor Leland Stanford of California and the acting governor of Nevada Territory, Orion Clemens (Mark Twain's older brother). The Houghton-Ives survey determined that Aurora, a mining town that was the seat of Mono County, actually lay three miles east of the California border. The team also determined that the intersection point at 120 degrees longitude and 39 degrees latitude fell in the middle of Lake Tahoe. From there they began surveying the oblique line angling southeast toward the Colorado River, but completed only slightly more than 100 miles of the line before winter blizzards ended the effort. More than 300 miles of that boundary remained unsurveyed.

After the Civil War, Daniel G. Major was surveying the Oregon-California boundary line for the General Land Office. He calculated that the intersection of that boundary with the 120th meridian actually lay more than two miles west of the mark established in 1863.

Consequently, another survey of the California-Nevada border line was conducted in 1872 by Allexey W. Von Schmidt. Though Von Schmidt was supposed to begin his north-south journey at Major's corrected point, he concluded that *neither* of the previously identified corners was correct and began between the two points. After following the longitude line into Lake Tahoe, he directed his survey toward the Colorado River. However, the river channel had shifted since 1863 by more than a mile. Von Schmidt moved his end point and then worked his way back northwest, resurveying about 270 miles of the line. He ended his correction where it intersected his own line from the north, even though that produced a kink in the eastern border

It took 20 years for Congress to appropriate funds, in 1892, for yet another survey of that troublesome border segment. The United States Coast and Geodetic Survey, in the next several years, marked a new, straight border line. Both states adopted the results by statute, California in 1901 and Nevada in 1903. Even so, the two states argued about the boundary until, in 1980, the issue went before the United States Supreme Court. The ruling was that the two states must live with the boundary as drawn.

The Other Grid System

After California achieved statehood, the United States Public Land Survey System (USPLSS) was applied there. This national mapping system established arbitrary starting places, called "initial points," from which lines heading east to west and north to south were surveyed. The north-south line running through an initial point was designated a "principal meridian." The east-west line through the point became a "baseline."

Working from those lines, surveyors divided public lands into townships (squares with six-mile sides), which were subdivided into 36 one-mile-square sections. Every

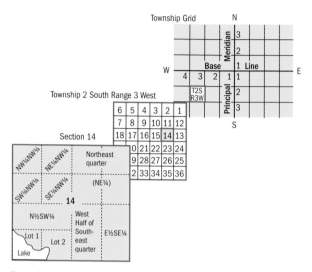

Figure 9. The public land survey system (PLSS) established range and township grids and numbered sections within each township. Section 14 shows both normal division of the section into aliquot parts and the fractional division into government lots.

township was identified with a township and range number. Locations north or south of the baseline were given a "township" number, and "range" designations indicated locations east or west of the principal meridian. The perpendicular regularity of many county roads and fence lines still reflects this imaginary grid imposed on the landscape (figure 9).

In 1851 Colonel Leander Ransom, a deputy surveyor with the General Land Office, placed a marker on the summit of Mount Diablo, east of San Francisco Bay, and used that landmark to survey prime meridians and baselines. Those lines became the basis for surveying most of California and Nevada. Although Mount Diablo does not sit at an exact latitude and longitude integer degree intersection (it is at 37°51'30" N and 121°54'48" W), its 3,845-foot

Plate 43. The vista, looking northeast, from the summit of Mount Diablo.

summit looms high above the nearby hills of the Coast Ranges. From the top, the Sierra Crest could be seen on clear days. More to the point, surveyors would be able to sight back toward the Mount Diablo marker from many miles away (pl. 43).

William Brewer surveyed the mountain in 1862. "From the peculiar figure of the country probably but few views in North America are more extensive—certainly nothing in Europe. . . . The great central valley of California . . . lies beneath us in all its great expanse for near or quite *three hundred miles of its length!* . . . I made an estimate from the map, based on the distances to known peaks, and found that the extent of land and sea embraced between the extreme limits of vision amounted to eighty thousand square miles" (Brewer, 1974 [1966], 264, 265, emphasis in original).

With Mount Diablo used as the initial point, the Sierra foothills town of Sonora occupies several sections in T1N, R14E, that is, tier 1 of townships north of the Mount Diablo baseline and column 14 of townships east of the principal meridian.

Because California angles across so much north-south and east-west territory, two other initial points became necessary within the state, one on San Bernardino Mountain in San Bernardino County (established in 1852) and the other on Mount Pierce in Humboldt County (in 1853) (map 6).

The new grid system facilitated ownership and transfers of land, which would thereafter take on new value as "real estate."

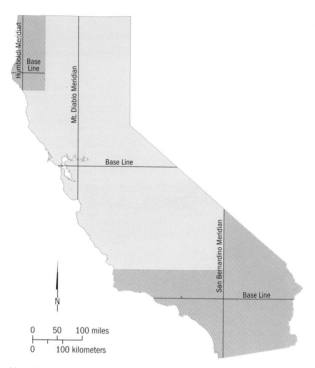

Map 6. Three baselines were needed to cover all of California, starting from Mount Diablo, San Bernardino Mountain, and Mount Pierce in Humboldt County.

Harvesting Wealth from the Earth

The '49ers first sought gold using shovels, pans, and small sluice boxes, working the riverbeds of the Mother Lode country in the western foothills of the Sierra Nevada (pl. 44). As the easiest pickings diminished, methods switched from placer mining to hardrock or lode mining—tunneling beneath the surface—and to hydraulic mining.

Minuscule bits of gold hiding in the earth were sought by hydraulic miners, who used powerful water cannons, called "monitors," to wash down entire hillsides and sluice the material to separate out the gold. Everything else was "waste"; all that had been alive in the soil, every plant that grew on the surface of those hills, and all the small creatures that tunneled through the ground were swept away and sent downstream. The practice began during the 1860s and peaked in the 1880s,

Plate 44. On the middle fork of the American River, in 1859, miners had turned the river out of its channel to seek gold in the gravels of the dry riverbed.

Plate 45. Eroded bluffs at Malakoff Diggings State Park mark where the powerful water cannons of hydraulic mining operations washed entire hillsides down through sluice boxes in the quest for gold.

when over 400 hydraulic mines were sending muddy wastes downriver (pl. 45). "[After] 1876 a hundred million cubic yards of gravel, sand, and clay . . . washed into the Yuba and its tributaries . . . in 1880 some 15,220 acres had been seriously injured" (Shinn, 1948 [1885], 255). The injury referred to was not the chewed-up landscape but the damage from "slickens" (fine silt) deposits on farmland and from floods that damaged valley towns after river bottoms filled with sediment and water escaped from river channels. In total, about 1.5 billion cubic yards of living earth became debris waste passed down onto Central Valley lands—eight times the volume of earth shifted during construction of the Panama Canal!

Finally, a Marysville land owner sued the owners of North Bloomfield Mine and other hydraulic mines operating on the Yuba River, seeking an injunction against the discharging of mining debris into rivers. California's first environmental impact lawsuit was decided by Judge Lorenzo Sawyer in January 1884, in favor of the valley farmers and townsfolk. Hydraulic mining was not banned outright by the Sawyer decision, but miners would no longer be allowed to injure downstream property interests. Material "wastes" thereafter had to be confined at mining sites. That requirement effectively ended hydraulic mining in California. The legal decision shocked late-nineteenth-century California society because the Gold Rush had driven the founding of the 34-year-old state and, since statehood, mining had enjoyed strong political favor.

The California Debris Commission was formed and oversaw the dredging of the lower Yuba River east of Marysville to mitigate the damage. Gravel was piled along the riverbanks, and in the twentieth century, mining companies reprocessed those deposits because they still held gold that the original sluices had not captured. Such reprocessing of tailing piles and former mine wastes became the principal source of gold in California in latter decades. After all the gold had been extracted along the Yuba River, the gravels were used as aggregate material for concrete (pl. 46).

"They Treated Soil Like Dirt"

(ANONYMOUS)

Early mining practices in California's gold country were codified in the new state's laws and served as the model for the

Plate 46. Braided debris washed down the Yuba River by hydraulic mining is still visible in this space shuttle photograph.

nation's 1872 Mining Law. Today that 138-year-old federal statute still controls mining activities on public lands. In the nineteenth century the government's goal was to populate the western states by transferring land ownership to miners (the Homestead Act and other land transfer grant laws also promoted that goal). Miners were required to pay no royalties on the minerals extracted from public lands. Control of the site was established by discovery and appropriation, with the requirement that the miner work the site to retain possession. Those conditions essentially remain part of present-day mining law, and federal legislation added the right to full and permanent ownership of the land (instead of mere possessory title) after a "patent" was issued by the government.

The U.S. Bureau of Land Management (BLM) administers mining claims on public lands. After filing a claim, miners must perform $100 worth of labor or improvements each year on their placer or lode claims. An annual

Plate 47. An open-pit mine in the Sierra Nevada.

filing is required, and owners of more than 10 claims pay a $125 maintenance fee; smaller mining operations receive fee waivers.

The 1872 Mining Law made no provision for environmental protection and did not address pollution. Other federal and state regulations have introduced water quality and land restoration requirements, but reform of the 1872 Mining Law has proven politically difficult.

The dominant practice in modern-day gold mining creates massive open-pit excavations to locate, as with hydraulic mining, minuscule pieces of gold within the earth. The material is heaped, and the ore is separated by leaching with liquid cyanide, which binds with gold (pl. 47). Stripping the topsoil and piling processed "wastes" for possible reclamation affects the physical, chemical, and biological properties of that soil. Deep in the piles, anaerobic conditions prevail. Populations of bacteria, fungi, actinomycetes, and other soil organisms decline, which makes later restoration difficult because of the reduced nutrient levels and lost soil ecosystem functions.

In 2007, $413 million worth of gold was produced by mining in California, according to the Department of Finance. Activity fluctuates in boom and bust cycles—a longtime characteristic of the mining industry that is related to variations in the price of gold.

Oil and natural gas are other products taken from the earth in California. The theory holds that fossil fuel creation began in the Carboniferous period, about 300 million years ago, before the time of the dinosaurs. Death and decomposition of plants in swamps and in shallow oceans, including algae and diatoms, formed carbon-rich peat soils, which when covered by sediments and rock gradually transformed into sedimentary rock. Pressure increased as covering deposits continued to build up, water was squeezed out, and eventually the fossil fuels formed. In most cases the deposits remained trapped in folds and pockets underground until tapped by drillers.

In 1850 General Andreas Pico collected oil from hand-dug pits in Los Angeles and refined it for use in lamps. The La Brea tar pits are a surface manifestation of oil in the region. In 1861 the first oil well was drilled in northern California, at Petrolia in Humboldt County. The first successful commercial oil field in the state was developed in 1875 in Pico Canyon in Los Angeles County. The Kern River oil field was discovered in the San Joaquin Valley in 1899, and this area would become the top oil-producing region in the state (pl. 48). But in the early 1920s, one of every five barrels of oil produced in the United States came out of the Los Angeles area, primarily from Huntington Beach, Santa Fe Springs, and the Long Beach (Signal Hill) oil fields. Today California ranks fourth in oil production (including offshore operations) among states, behind Louisiana, Texas, and Alaska. Yet the state's voracious demand for energy—particularly oil to fuel transportation—makes it a net importer of oil, mostly from foreign nations.

Plate 48. Oil rigs in Kern County.

"Land" Becomes "Real Estate"

Train tracks tied Sacramento to the rest of the nation after the golden spike was driven at Promontory Point, Utah, in 1869. The new transportation system changed many things, but it especially influenced land ownership and development patterns within California. The Pacific Railway Act of July 1, 1862, granted title to alternating sections of public land for 10 miles on either side of the track right-of-way to the railroad companies. Later, the zone on either side of the track was expanded to 20 miles. For every mile of rails they laid, the companies acquired up to 12,800 acres of federal land. Railroad land grants in California ultimately totaled 11,585,534 acres—more than 11 percent of California's total land acreage (map 7). That land wealth bought potent political power for the railroads, particularly Central Pacific Railroad Company, which later became Southern Pacific (SP), the sole rail provider to the state from 1869 to 1887. Railroad companies set out to profit from their holdings by enticing settlers, who would grow crops to be shipped east by railroad, and by encouraging tourism to "the Golden State."

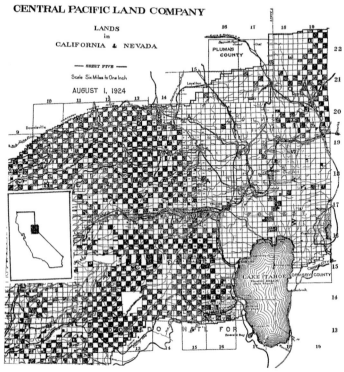

Map 7. Detail from a map of alternating land sections granted to the Central Pacific Railroad near Lake Tahoe, 1924.

San Francisco was the leading California city for many decades after the discovery of gold and the declaration of statehood. The population of Los Angeles stood at 11,183 in 1880. A real estate "boom" in the 1880s expanded the southern California city to 50,000 residents. Land syndicates created a hundred new communities offering 500,000 subdivision lots. Construction never took place on many of them. The boom really took hold when SP's competitor—the Atchison, Topeka and Santa Fe Railroad—pushed its

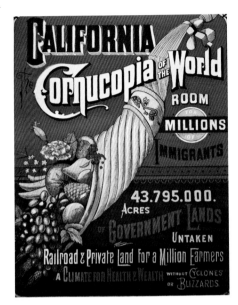

Plate 49. A California Immigration Commission poster advertising available land in California in 1883.

track into southern California in 1887. A fare "rate war" between the Santa Fe and SP allowed a person to travel from the Missouri River to California for only one dollar! Return fares, however, were not so cheap. Railroad company advertising saturated the eastern states and even Europe, promoting California and its tourism, healthy climate, agriculture, and available land for immigrants (pl. 49).

Books, pamphlets, and posters extolled "semi-tropical" southern California and "America's Italy." Major Benjamin C. Truman, chief of SP's literary bureau, authored *Semi-Tropical California: Its Climate, Healthfulness, Productiveness and Scenery* in 1874. Perhaps the most widely read publication promoted by SP was journalist Charles Nordhoff's *California for Health, Pleasure and Residence.* "In the great San Joaquin Valley," Nordhoff wrote, "which is just being opened by the building of the Southern Pacific Railroad, there are three millions of acres open to settlement. . . . There Government land can be had in eighty-acre tracts, under the Homestead Act, for nothing, and one hundred and sixty acres at . . . two dollars and a half per acre; while the railroad sections can be bought at low prices, and on five years' credit, in whole sections of six hundred and forty acres. If any . . . person in the East . . . should ask me what I would advise . . . I should . . . direct him first to . . . the land-office of the Central Pacific Railroad Company" (Nordhoff, 1873, 177–178).

Southern Pacific was not the only one with a marketing campaign. The Santa Fe Railroad created the California Excursion Association. In 1889 the Merchants and Manufacturers Association was formed with the backing of *Los Angeles Times* publisher Harrison Gray Otis. *Sunset* magazine was founded by SP to promote westward travel. The name *Sunset* came from the Sunset Limited train, which ran between New Orleans and Los Angeles.

Some locals even then were wary of urban expansion onto prime farmland. Mary C. Vail wrote *Both Sides Told, or Southern California as It Is . . .* , an 1888 pamphlet, where she expressed her dismay that "a wonderful extent of these fertile valleys are staked off in town lots. It certainly would have been much better if these expectant townsites were yet under the direction of the intelligent plowman" (1888, 8). "All around the growing town-centers, land is too valuable to be devoted to farming" (16). That pattern, real estate pressures forcing farmland out of production, has been a challenge for the state ever since.

Citrus orchards had been present since the Mission days, but in 1873 two navel orange trees, a new variety, were planted in Riverside. A few years later the Valencia orange was introduced to Orange County. The two varieties were suited to different micro-climates of the coastal plains and foothills and, since they ripened in different seasons, were not direct competitors.

Citrus farming's success was aided by partnership with the railroads, building on their expertise at marketing. Citrus crate

Plate 50. Citrus labels advertised the fruit inside packing crates, but also painted an idyllic picture of life in California in the early twentieth century.

labels became a new art medium. Although each citrus brand had a unique label, most depicted bounteous harvests and sun-drenched vistas in warm, lush colors. The labels marketed the fruit, but also an Eden-like image of the California landscape (pl. 50).

The emerging citrus industry brought government pomologist G. Harold Powell to southern California in 1904 and 1909. Powell's letters to his wife detailed a quality of life in Los Angeles and its rural surroundings that sounds enormously attractive today: "This morning I was out at 6:30 and went to Hollywood again to look over lemon groves. Los Angeles is one of the prettiest cities I have seen. . . . The city is a hustling business town, over 100,000 people, fine blocks, elegant hotels, and real estate agents thick enough to walk on" (Powell, 1990, 20, 21).

For over 70 years citrus groves provided beauty and wealth in southern California, until their fertile soils were covered by housing tracts, concrete, and asphalt. Los Angeles County once had some of the most productive farm soils in the nation, but agricultural acreage fell from 300,000 to 10,000 acres between World War II and 1970.

There is value in any experience that reminds us of our dependency on the soil-plant-animal-man food chain, and on the fundamental organization of the biota. Civilization has so cluttered this elemental man-earth relation with gadgets and middlemen that awareness of it is growing dim. We fancy that industry supports us, forgetting what supports industry.

LEOPOLD (1974 [1949], 212)

MORE THAN HALF of California's acreage is considered wildland, land that has not succumbed to urbanization or become farmland. The living soil performs its essential functions in most forested land, rangeland, and wilderness tracts. Those processes are left to work naturally unless wildland soils have been eroded or destabilized by human activities. Where timber is harvested or mining occurs, land managers may have to act to restore or minimize damage to soils. Erosion from dirt roads and off-road vehicle tracks through wildlands is the major source of sediment runoff into the waters of the Sierra Nevada and Coast Ranges, which makes informed decisions about road building and vehicle access within these lands critically important.

Healthy soil is essential to productive forests and wildlands. Modern approaches to promote biodiversity aim not just at keeping endangered species from going extinct, but also at maintaining the complexity of the ecosystem and the redundancy required for it to continue to function. At the soil level, the nitrification function, with multiple populations of microbes converting ammonium to nitrate, can continue even if one species dies out. Biodiversity becomes insurance for ecosystems.

Protective measures for open space and wildlands usually focus on what is observed on the land surface, but as with all human impacts on California's environment, the soil is the fundamental basis for wildland health. This principle can be difficult to honor properly as communities debate the importance of open space versus development or as controversial practices are undertaken on protected lands.

Timber harvest options provide an example of this tension, where impacts on soil health are not always obvious. Methods used for cutting timber and hauling it off-site, and the choice between clear-cutting and selective harvesting, affect the magnitude of site disturbance. The amount of slash—the smaller limbs left after lumber-size wood has been removed—and the techniques chosen to reduce the volume

Plate 51. A clear-cutting timber operation in the Stanislaus National Forest.

of slash as a fuel-load on the ground, have consequences for soil organisms. Tractors, wheeled skidders, and mechanized harvesting equipment disturb the soil surface and compact the soil along skid trails and logging roads (pl. 51). Compaction can reduce soil organism activity and promote erosion. The use of selective harvest techniques, such as helicopter logging, can minimize those impacts.

Forestry agencies and responsible logging companies can take protective steps such as inoculating new tree seedlings with mycorrhizal fungi before planting and retaining patches of intact forest among harvested areas to serve as sources of soil organisms that can vitalize the new growth. Downed logs, left in place, send out mycorrhizal hyphae through the soil and serve as sources of nitrogen-fixing bacteria and other microbes, and of arthropods and small mammals. Originating from large rotting logs, fungi can reconnect the disturbed forest grid and send water and nutrients to growing seedlings. Downed logs on the surface also slow erosion and increase water infiltration (pl. 52).

Plate 52. A decomposing log is home to many organisms, including flowering plants.

The trees in wildland forests are part of broader vegetation communities. Researchers at the University of California at Davis recently learned that blue oak trees create enhanced soil fertility "islands" beneath their canopies. They found that the ground below the trees had more organic matter and available nutrients than adjacent grassland in the oak woodlands and savannas. Removing oak trees led to rapid deterioration of soil quality.

Fire and Soil

The effects of fire on soils were explained in *Introduction to Fire in California*, one of the other titles in the Californians and Their Environment series. Plant nutrients locked up in standing vegetation, dead or alive, are recycled by fires (pl. 53). Mineral soil exposed after burning is a prerequisite

Plate 53. Ash residues left after fires burn become part of the nutrients released to the soil.

for successful seed germination in some species. Giant sequoias, like Coulter pines and a number of other trees with serotinous cones, hold back on the release of seeds until after a fire, to take advantage of the prepared ground. The big trees cannot successfully reproduce in the absence of regular, low-intensity ground fires.

During low-intensity fires, only the upper layer of organic material may burn, with little of the heat penetrating downward. Intensely hot fires, however, can consume organic matter and fungi in the upper soil layers. Countering this action is the fertilization effect from the release of nutrients that had been stored in vegetation. Fires may make a flush of nitrogen available for plants, but hotter fires can also volatilize nitrogen, causing it to evaporate into the air and leaving less in the ground. The nutrients phosphorus and calcium are not as easily volatilized and become available to plants as ash is reincorporated into the soil.

Extremely intense fires, including many that burn in chaparral shrublands, sometimes increase soil water repellency. The result is a "hydrophobic soil," where water is not absorbed well by the ground. The layer forms because heat activates resins common in chaparral shrubs, which settle onto soil particles. Water repellency is temporary. The resin substances are slightly water soluble, and burrowing animals and growing roots gradually break apart the band in the soil where the chemicals settled. While the effect persists, it may increase runoff and contribute to erosion or debris flows.

Heavy rains falling onto hillsides where surface vegetation and ground litter have been burned off can generate floods that pose dangers for humans in the watershed (pl. 54). On Christmas Day 2003, two months after major fires burned in the San Bernardino Mountains, torrential rains generated a 20-foot-thick flow of mud and debris that swept into a youth camp. Fifteen people died.

Federal land management agencies are required to conduct an evaluation after a fire and to view the situation as an

Plate 54. A post-fire debris flow.

emergency until necessary rehabilitation measures have been taken. Teams focus on threats to life and property, erosion control, water quality, and loss of soil productivity.

Wildland Management

California has an array of wildland management agencies whose different objectives and policies can lead to confusion. Key differences matter; navigating through that confusion can help citizens realistically judge among the agencies' options and assess how well they are meeting their legal mandates. A key difference is agencies whose primary purpose is resource preservation, as in state and national parks, versus those with "multiple use" requirements, such as state and national forests. Wildlife management refuges must balance conservation and habitat preservation with recreational hunting opportunities. Other regional, local, and nongovernmental land management agencies and organizations also control and preserve significant amounts of California's wild acreage.

Nearly half of the land in California, almost 48 million acres, is federal land or Indian land held as a federal trust. Only Alaska and Nevada have a greater percentage of state acreage in the form of federal lands. The federal government makes "payments in lieu of taxes" to counties to offset the loss of tax revenues because of these tax-exempt lands. California counties received $21 million in payments in the year 2000. Other revenues, generated by mineral leases, livestock grazing, and timber sales, are also shared.

Significant amounts of ranch lands that provide grazing for livestock also provide wildlife habitat and open-space relief from urban congestion. Most are remnants of the Spanish and Mexican land grants that placed large holdings of land in the hands of a few Californians. Where such lands

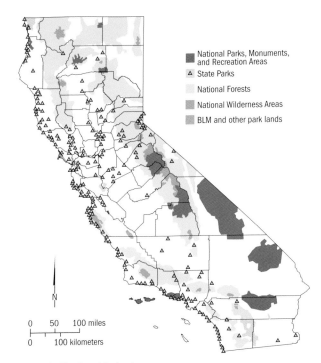

Map 8. California public lands.

National Parks, Monuments, and Recreation Areas
△ State Parks
National Forests
National Wilderness Areas
BLM and other park lands

0 50 100 miles
0   100 kilometers

N

have escaped development, efforts to conserve large parcels that retain wildland qualities have become increasingly important (map 8).

The story of park lands in California begins not with federal land management, but with an 1864 federal grant to the young state.

"Protect, Restore, Maintain, and Sustain"

(FROM THE CALIFORNIA STATE PARKS MISSION STATEMENT)

President Abraham Lincoln and the nation were dealing with the third year of the Civil War when Lincoln signed legislation establishing the Yosemite Grant. Twenty thousand acres including Yosemite Valley and the Mariposa Big Trees Grove were transferred to the state government. Grants of federal land to states were not particularly unusual, but the express conditions in this case established a precedent in public land preservation. The land was to "be held for public use, resort, and recreation, and . . . be inalienable for all time."

In 1890 another 900,000 acres of wildlands surrounding Yosemite Valley were given federal protection, an effort pushed by John Muir and the newly formed Sierra Club. In 1906, after 42 years as a California park, Yosemite Valley and the giant sequoia grove were returned by the state to federal government control and added to Yosemite National Park.

The earliest park unit that is still a part of the California State Parks system was created while Yosemite Valley was under state control. In 1891 the Marshall Monument was established to commemorate the discovery of gold by James Marshall on the south fork of the American River. In 1902 the state's first wildland park in the modern system, California Redwood Park, was created in the mountains north of Santa Cruz. It was later renamed Big Basin Redwoods State Park (pl. 55).

Today the California State Park System manages 278 park units on 1.5 million acres and along 320 miles of the Pacific coast. It hosted 77 million visitors in 2005. Yet in 2009, as this book was being prepared, state agencies faced massive budget cuts and more than 100 state park units were threatened with closure.

Some parks feature resources and unique sights directly tied to local soil conditions. The isolation of coast live oak groves at Los Osos Oaks State Reserve near San Luis Obispo reflects edaphic soil conditions. Relationships between soil and vegetation are revealed most clearly on the "ecological staircase" trail at the Jughandle State Natural Reserve along the Mendocino coast. There visitors ascend inland through five successive terraces, each uplifted from its origin at sea

Plate 55. Big Basin State Redwood Park in the 1920s, with Assistant Warden Joseph P. Park patrolling. A small sign to the left of the vehicle reads, "Do Not Pick Flowers or Shrubs" and "Do Not Build Fires."

level and each representing about 100,000 years of time and 100 feet gain in elevation. The terraces' sandstone cliff faces were sculpted by ocean waves, and then the land rose—about an inch a century—as the result of pressures generated as the Pacific and North American tectonic plates collided. Weathering and leaching produced increasingly nutrient-poor, highly acidic conditions, which are reflected in the plant communities. Prairie grasses and Sitka spruce grow at sea level. The next level holds a pine forest, including coast redwoods. Upper terraces support a unique "pygmy forest" where mature trees remain small in the ancient, nutrient-depleted soils (pl. 56). The process is similar to the nutrient deprivation purposefully applied to produce dwarf plants in Japanese bonsai gardens.

One lesson apparent at the Jughandle State Natural Reserve is that natural processes do not inevitably improve or

Plate 56. At Jughandle State Reserve, a state park ranger marks the height of a Mendocino pygmy cypress *(Cupressus goveniana* subsp. *pygmaea)*, full grown yet only 4 feet tall. The taller pines *(Pinus contorta* subsp. *bolanderi)* in the background are also much shorter than beach pines growing in more nutrient-rich soils.

even sustain soil fertility. Yet pygmy forests and ancient soils that become severely depleted of nutrients are not common, because soil-building forces, such as glaciation, volcanic eruptions, sedimentation, and wind deposition, usually counter soil depletion processes.

"Habitats upon Which They Depend"

(FROM CALIFORNIA DEPARTMENT OF FISH & GAME MISSION STATEMENT)

The other state agencies managing large tracts of wildland are the Department of Fish & Game and CALFIRE (otherwise called the California Department of Forestry and Fire Protection). Fish & Game manages over 900,000 acres of land and water resources, wildlife areas, ecological reserves, and hatcheries. CALFIRE's primary purpose is to control wildland fire, but the agency also administers eight demonstration state forests on 71,000 acres.

"Unimpaired for . . . Future Generations"

(NATIONAL PARK SERVICE ORGANIC ACT)

Over 8.2 million acres are preserved in California by the National Park Service. This management presence in California dates back to 1890, when lands surrounding Yosemite Valley were set aside and, in the same legislation, Sequoia and Grant Grove national parks were created in the southern Sierra Nevada. Later additions created the much larger Sequoia and Kings Canyon national parks in 1940 (pl. 57).

The National Park System includes both national parks and national monuments. Parks are established by Congress to conserve outstanding scenic features or natural phenomena with inspirational, educational, or recreational value, and to ensure they are left "unimpaired for the enjoyment of future generations." By comparison, national monuments

Plate 57. The east entrance to Yosemite National Park is at Tioga Pass at the crest of the Sierra Nevada.

preserve lands with historic, prehistoric, or scientific interest. They can be established by presidential proclamation and then approved by Congress, or occasionally by a direct act of Congress.

California also has two very large national recreation areas (NRAs) administered by the National Park Service, Golden Gate NRA and Santa Monica Mountains NRA (table 2).

TABLE 2. National Park System Land Management Areas in California

	Total Acres	Wilderness Acres
National Parks		
Channel Islands	249,561	–
Death Valley	3,372,402	–
Joshua Tree	789,746	429,690
Kings Canyon	461,901	456,552
Lassen Volcanic	106,392	78,982
Redwood	112,512	–
Sequoia	404,051	280,428
Yosemite	761,266	677,600
National Monuments		
Cabrillo	160	–
Devils Postpile	799	750
Lava Beds	46,560	28,460
Muir Woods	554	–
Pinnacles	24,514	16,048
National Seashore		
Point Reyes	71,068	25,370
National Preserve		
Mojave	1,534,819	–
National Recreation Areas		
Golden Gate	74,820	–
Santa Monica Mountains	154,095	–
Whiskeytown-Shasta-Trinity	42,504	–

NOTE: The National Park Service also administers four National Historic Sites and two National Historic Parks in California.

Plate 58. Volunteers doing trail work in the Golden Gate National Recreation Area.

These "urban national parks" protect open-space lands and historic resources while serving large regional populations. The Golden Gate NRA has nearly 20 million visitors a year and includes ocean beaches, redwood forests, lagoons, marshes, military properties, and Alcatraz Island (pl. 58). The Santa Monica Mountains NRA provides 580 miles of hiking trails and wildland, relieving the otherwise nearly unbroken expanse of urbanization spreading across the Los Angeles basin.

"Land of Many Uses"

(USDA FOREST SERVICE SLOGAN)

The National Forest Reserve Act of 1891 gave the president authority to set apart forest reserves on public lands for the conservation of natural resources. Forest reserves were originally administered by the Department of the Interior. The Division of Forestry, created in 1899, was moved to the Department of Agriculture, and in 1905 the

USDA National Forest Service was established. Inclusion in the Department of Agriculture rather than the Interior Department reflected a key difference between the forest service's and park system's operating goals. Forests are managed for "multiple use," including both protection and extraction of forest and mineral resources, and to provide livestock grazing land, hunting, and public recreation. Mission definition is a constant challenge for an agency like the National Forest Service, which must balance so many competing uses.

In California there are 20 national forests incorporating over 20.6 million acres. Nine national forests (and the Tahoe Basin Management Unit) control 40 percent of the land in the Sierra Nevada (pl. 59). Southern and central California have the Angeles, San Bernardino, Cleveland, and Los Padres national forests, which present tremendous challenges related to their proximity to millions of California residents. In the north of the state are the Mendocino, Modoc, Rogue River, Klamath, and Shasta-Trinity national forests.

"Management, Protection, Development and Enhancement"

(FEDERAL LAND POLICY AND MANAGEMENT ACT OF 1976)

The Bureau of Land Management (BLM) administers other public lands, including 15.2 million acres in California. BLM is also a multiple-use agency and has particular responsibility for the "subsurface mineral estate," that is, the mining lands and mining activities on federal lands, underlying privately owned lands, and on Native American tribal lands (592,000 acres in California).

The U.S. government's General Land Office and the Grazing Service were combined in 1945 to form the Bureau

of Land Management. From the colonial era through the late 1800s, over 1 billion acres of federal land were transferred to state or private ownership, much of that by the General Land Office (established in 1812). In 1976 national policy changed so that federal lands would be retained unless disposing of particular parcels served the national interest.

BLM administers the 250,000-acre Carrizo Plains National Monument, the Santa Rosa and San Jacinto Mountains National Monument, and the California Coastal National Monument, which includes more than 20,000 small islands, rocks, exposed reefs, and pinnacles in a 12-mile-wide offshore corridor running the length of the state.

"Understand, Appreciate, and Wisely Use"

(FROM U.S. FISH & WILDLIFE SERVICE OBJECTIVES)

The United States Fish & Wildlife Service (USFWS) protects almost 80,000 acres in California, including 37 national wildlife refuges and wildlife management areas (also two fish hatcheries). About half of the refuges are open to the public for outdoor recreation. The Sacramento National Wildlife Refuge Complex includes five wildlife refuges and three wildlife management areas in the Sacramento Valley totaling over 35,000 acres of wetlands and uplands that serve nearly half the migratory birds on the Pacific Flyway (pl. 60).

Other USFWS lands are closed to protect wildlife habitat or for safety reasons. For example, Marin Islands National Wildlife Refuge consists of two small islands in San Pablo Bay off the coast of San Rafael. They are rookeries for herons and egrets, and support many other bird species. The islands

Plate 60. Don Edwards San Francisco Bay National Wildlife Refuge.

are closed to the public because of their dangerous, sheer cliffs and to avoid disturbing the bird populations.

"Stewardship Responsibilities"

(FROM DEPARTMENT OF DEFENSE ENVIRONMENTAL CONSERVATION PROGRAM DEFINITIONS)

There are about 4 million acres of military land in California. The Department of Defense is required to manage its natural resources and, where possible, public access is allowed for outdoor recreation (although military requirements take precedence). After the Cold War, 29 bases in California were put on a closure list. Parts of the Fort Ord army base near Monterey Bay have recently been added to the state park system (pl. 61). Where sand dunes were once used to stop bullets on a firing range, thousands of pounds of lead-contaminated soil had to be hauled away before public access was allowed. The El Toro Marine Corps Air Station in Orange County

Plate 61. At the Ford Ord Dunes State Park, rifle ranges were cleansed of thousands of pounds of toxic lead left by spent ammunition before the area was opened to the public.

was slated to become a regional airport after the base closed in 1999, but voters instead chose to create a new Orange County Central Park and Nature Preserve there, managed in a partnership by the City of Irvine, the federal government, and a private landowner, with rights to develop a planned community along with 1,300 acres of public space.

Despite the maneuvers and explosives that go along with military land use, large parcels of open space such as Camp Pendleton, along the coast between San Diego and Orange County, serve as wildlife habitat and provide corridors for wildlife movement. The Army Corps of Engineers also administers 22 park and recreation areas in the state.

"Where Man Himself Is a Visitor"

(FROM THE WILDERNESS ACT OF 1964)

California has 148 designated wilderness areas, the greatest number of any state (though Alaska has much more total wilderness acreage). California's wildernesses are on land that meets the Wilderness Act of 1964 criteria inside national parks, monuments, and forests, and also on public lands administered by BLM and by the USFWS. The state's wilderness acreage expanded to 14,963,867 acres with the addition of over 700,000 acres in March 2009, when the president signed federal legislation that made additions to 10 existing wildernesses and created 10 new wilderness areas in the eastern Sierra, the San Gabriel Mountains, Riverside County, and within Sequoia–Kings Canyon National Park.

According to the Wilderness Act, these areas are designated "to preserve the unique wild and scenic areas of America's public lands." The act defined wilderness as "an area where the earth and its community of life are untrammeled by man, where man himself is a visitor who does not remain," and as land retaining its primeval character and influence, without permanent improvements or human habitation, which is protected and managed so as to preserve its natural conditions and which (1) generally appears to have been affected primarily by the forces of nature, with the imprint of man's work substantially unnoticeable; (2) has outstanding opportunities for solitude or a primitive and unconfined type of recreation; (3) has at least five thousand acres of land or is of sufficient size as to make practicable its preservation and use in an unimpaired condition; and (4) may also contain ecological, geological, or other features of scientific, educational, scenic, or historical value.

Misconceptions abound concerning what is and what is not allowed in wilderness. Opponents regularly claim that the designation "locks people out" or that the only way to travel in wilderness is as a hiker. In fact, over 12 million people visit the nation's wilderness areas each year to ride horses, hunt game, fish, climb mountains, cross-country ski, raft or canoe on rivers and lakes, as well as to hike. Recreation uses that rely on mechanical transport or motorized equipment are not permitted (exceptions include wheelchairs and, in Alaska, certain mechanized and motorized means associated with traditional and subsistence activities).

"Embrace Life!"

(EAST BAY REGIONAL PARKS SLOGAN)

City, county, and regional governments provide open space and outdoor recreation services to most Californians, though their land holdings totaled less than 600,000 acres of park lands in 2002, according to the California Department of Parks and Recreation. Cities and counties are authorized by

Plate 62. The East Bay Regional Parks District celebrated its 75th anniversary in 2009 at the Crab Cove facility.

state law to require developers to set aside land, donate conservation easements, or pay cash in lieu of land and/or impact fees. Revenues generated go toward park improvements

Interest in the outdoors explains why, in 1934, amid the Great Depression, voters taxed themselves to acquire open space in the East Bay hills. The East Bay Regional Parks District grew from initial land acquisitions in the watershed above Oakland and Berkeley, to today's 65 regional parks, 31 regional trails, 1,000 miles of park trails, and 98,000 acres that serve 14 million annual visitors (pl. 62).

A California Park and Recreation Society survey in 2009 found that 98 percent of households—almost all Californians—visit or participate in park programs. Those surveyed placed the highest value and priority on preservation of, and access to, outdoor spaces, especially places that have been minimally developed, are in a nearly natural state, or offer facilities for play, exercise, and group sports (California Park and Recreation Society, 2009).

Another outdoor recreation survey was completed in 2005 for the state parks system. When asked to identify factors that were very important to their enjoyment of the outdoors, more than half of those surveyed chose being able to relax; feeling safe and secure; being in the outdoors; the beauty of the area; getting away from crowded situations; releasing or reducing tension; the quality of the natural setting; being with family and friends; doing something that children enjoy; having a change from the daily routine; and keeping fit and healthy (California Department of Parks and Recreation, 2005).

As positive (from the perspective of parks and recreation professionals) as these survey results seem, there has been another, more disturbing trend: the children of California and the rest of this nation seem increasingly disconnected from nature. The outdoors holds little interest for many young people, whose time and attention have been captured by computer games and electronic media that work best indoors, where there are handy electrical outlets. Even outside, headphones become effective blockades against the sounds of the real world. The perception that there is little of interest and nothing much to do outside inspired Richard Louv's book *Last Child in the Woods: Saving Our Children from Nature Deficit Disorder* and "Leave No Child Inside" campaigns. Nature deficit disorder is widespread, though thankfully not ubiquitous. The problem fundamentally lies not with children, but with the lack of outdoor experiences available to them and the values exhibited by parents and adult role models.

"A Beautiful, Restored, and Accessible Coastline"

(FROM COASTAL CONSERVANCY MISSION STATEMENT)

A number of independent state conservancies come under the umbrella of the California Resources Agency. In 2004 the Sierra Nevada Conservancy was created with conservation efforts covering the length of the mountain range, across 25 million acres in 22 counties. As another example, the California Coastal Conservancy purchases, protects, and restores coastal wetlands and provides for public access to California's shoreline.

The California Coastal Commission is an entity distinct from the California Coastal Conservancy. In 1972 Californians concerned about private development blocking access to the shoreline passed the "Save Our Coast" initiative, which created the California Coastal Commission, with jurisdiction over land use decisions in the coastal zone. In 1976 the legislature made the Coastal Commission a permanent, quasi-judicial state agency with authority to plan and regulate development and natural resource use along the coast in partnership with local governments.

The Coastal Commission also has authority to review activities in the coastal zone to protect shoreline access, scenic landscapes, and views of the sea; protect and restore sensitive habitats and natural landforms; and also protect archaeological sites, farmland, commercial fisheries, and special communities in the coastal zone. Their zone of influence reaches from three miles out at sea to an inland area whose width varies from a few blocks in urban areas to several miles in less developed regions.

While development projects in the coastal zone require permits (from the commission or the appropriate local government), some actions are exempt, including many repairs and improvements to single-family homes, certain "temporary events," and replacement of structures destroyed by natural disaster.

Control over development and uses of private land has often been controversial. Yet where public access and recreation opportunities on the coast remain available, where wetlands have not been filled, where views toward the sea are not blocked, and where farming still occurs on coastal

Plate 63. Footprints to the beach. A sign marks a coastal access corridor that is hard to discover among nearby residences.

terraces, the Coastal Commission has likely been doing its work (pl. 63).

"Legacy Projects"
(NATURE CONSERVANCY LAND PROTECTION CATEGORY)

At the beginning of the twenty-first century, about 120 land trusts were active in California. The Trust for Public Land, the Nature Conservancy, and the Save the Redwoods League are the largest organizations of this type working in California. They have used their fundraising abilities to acquire about 250,000 acres in California and protect another 91,000 acres with conservation easements, where landowners agree to keep their property undeveloped in return for tax benefits. Another 230,000 acres have been purchased and then transferred to the state or federal government.

As one example, the Nature Conservancy's mission is to preserve the plants, animals, and natural communities that represent the diversity of life on Earth by protecting the lands

Plate 64. Elkhorn Slough, near Moss Landing. A part of the research reserve was saved by the Nature Conservancy.

and waters they need to survive. Founded in 1958, the Nature Conservancy has worked on more than 100 projects in California and manages its own preserve lands. The Nature Conservancy refers to land transfers made to local conservation organizations or public agencies as its "legacy projects" or preserves. Such efforts helped establish the Carrizo Plain National Monument, Elkhorn Slough Reserve (pl. 64), and Guadalupe-Nipomo Dunes.

Who Owns All That Ranch Land?

Driving through rural areas, particularly the foothills of the state's mountain ranges, Californians might wonder, "Who owns all that ranch land?" The answer often goes all the way back to Spanish- and Mexican-era land grants, when major tracts became the property of a relatively small number of individuals. Many of those land grants, particularly on the flatter coastal plains, have been sold in parcels and developed. Those sections that still exist as open space often represent increasingly valuable wildland habitat for species that

Plate 65. California Condors feeding. Scavengers are part of the "recycling crew" that promotes new life following death.

elsewhere have seen their ranges shrink along with their ability to survive. The California Condor (pl. 65) is one of the most celebrated examples of a highly endangered species whose survival depends today on the availability of historic ranch lands.

In 1998, when the Hearst Corporation tried to build a resort complex with hotels and a golf course near the famous Hearst Castle along the central California coast, the California Coastal Commission found the plan contrary to its mandate to protect the coastal zone environment. Negotiations with the state of California led to a conservation easement announced in 2005. The state paid the Hearst Corporation $95 million to retire development rights and protect ranching and wildlife habitat on 80,000 acres east of Highway 1. Thirteen miles of shoreline along that scenic stretch of the Coast Highway were transferred to the California State Park System. A new 18-mile segment of the California Coastal Trail will be included. A section of Highway 1 that was endangered by eroding bluffs was scheduled to be moved east onto former Hearst Ranch land. In return, the Hearst

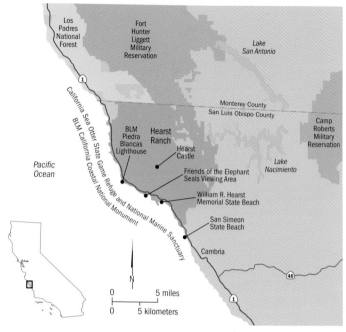

Map 9. Hearst Ranch.

Corporation would be allowed to build a 100-room hotel and a number of homes east of the highway (map 9).

An agreement affecting much more land, further south, was reached in 2008 to preserve 240,000 acres of the Tejon Ranch, in the hills north of Los Angeles and stretching northeastward to the southern Sierra Nevada (map 10). Rancho El Tejon's history goes back to 1843 and a Mexican land grant. The ranch was purchased in segments, beginning in 1855, by Edward Fitzgerald Beale. The creation of Tejon Ranch Conservancy to preserve 90 percent of the ranchland as open space, while allowing some development along with existing buildings and historic uses (cattle grazing and moviemaking), was negotiated by a coalition

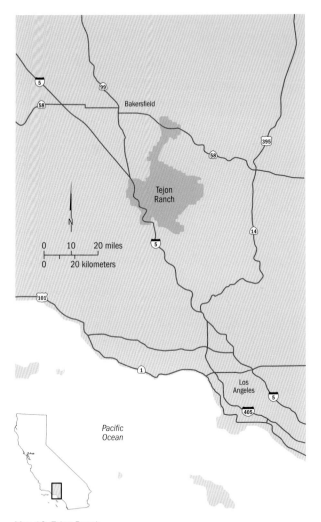

Map 10. Tejon Ranch.

of environmental groups including the Natural Resources Defense Council, the Sierra Club, and Audubon Society of California.

Tejon Ranch land provides habitat for California Condors and the endangered San Joaquin Kit Fox. It incorporates an intriguing meeting point for four ecological regions: the Sierra Nevada, Mojave Desert, San Joaquin Valley, and Coast Ranges. Thirty-seven miles of the Pacific Crest Trail will be re-routed out of the Mojave Desert as ranch land becomes accessible to the public. Some of the ranch land is to become a state park.

In return for this wildland preservation, environmental groups agreed not to contest plans for development on the other 10 percent of the ranch. The proposed development is significant and not without controversy, because it includes 23,000 homes with 80,000 residents along the ridge slopes and an industrial-commercial center near the Interstate 5 corridor. Gravel mining and oil and gas extraction activities also will be allowed within defined areas where they have been occurring and will be allowed some expansion. The Tejon Ranch Conservancy, formed to oversee the agreement and development as projects move forward, will require users of the land, including hunters and gravel miners, to use "environmentally friendly techniques." For example, hunters will not be allowed to use lead ammunition, since lead in carcasses has been identified as a threat to condors.

On May 14, 2008, the *Los Angeles Times* included an editorial headlined, "Environmentalists accepted development in exchange for land. But their work isn't done." The editorial pointed out that by "adding nearly 80,000 new residents to the far reaches of the Los Angeles region, the Tejon Ranch plan exemplifies sprawl, with all the attendant concerns about water, traffic, air quality and fire risks. These potential problems cannot be overlooked, no matter how much land is conserved."

An earlier, exceedingly controversial proposal to develop Newhall Ranch, on the north slopes of the Santa Susana mountains, did not include the ameliorating conservation easement aspects of the Tejon Ranch agreement. In the 1870s Henry Mayo Newhall acquired five Mexican land grant ranchos totaling 143,000 acres. In 1999 the proposed Newhall Ranch subdivision, at a similar scale to the Tejon development proposals, gained the approval of Los Angeles County. That plan included preparations for 21,000 houses and about 70,000 residents. Concerns about water supply and downstream impacts on the Santa Clara River, along with biological impacts, generated a series of lawsuits, public hearings, and further studies. In 2004 the Newhall Ranch Specific Plan was ruled compliant with state law and fully approved.

However, with work scheduled to begin on the first phase of construction, the Newhall Ranch corporation declared bankruptcy in 2009. A key creditor was pressuring the developer to sell the land to repay over a billion dollars in loans.

By the early twenty-first century, California was dealing with the consequences of two centuries of land development and population growth that had been accelerating ever since the Gold Rush and statehood in 1850. The state had become the epicenter for threatened and endangered species. California was considered a "biodiversity hotspot" on the planet because of its high number of at-risk endemics (species found only there).

It should be no surprise that efforts to further develop any remaining wildlands are met with dismay and organized opposition. Urbanization, habitat encroachment, pollution, large-scale agriculture, mining and oil extraction, invasive alien species, logging, increasingly severe wildfires, and—driving all of these forces—the pressures of human population growth threaten wildlands while increasing the intrinsic value of what remains. The soil seems helpless to resist the pressures we apply, and yet, now and then, the Earth sends us a jolt, reminding us that the land itself in truth cannot be tamed.

UNTAMED LAND

Earthquakes

Fourteen days north of San Diego, a 1769 expedition led by Don Gaspar de Portola reached a broad, tree-lined stream. Father Juan Crespi, diarist for the expedition, named it "the River of the Sweet Name of Jesus of the Earthquakes," because "we experienced here a horrifying earthquake, which was repeated four times during the day" (Bolton, 1927, 142). The plain through which the river ran they named "the Valley of Saint Anne," or Santa Ana. As they continued northward, leaving what would become Orange County and entering the Los Angeles basin, more quakes regularly jarred the landscape and their nerves.

California is notorious as a land of natural hazards, where the ground shakes and slides, wildfires rage, and at long intervals, volcanoes erupt; where "real estate" periodically gives the residents a reality check about humanity's being in charge of this landscape. That phrase, "natural hazards," is interesting, since natural processes on the land become "hazards" only when humans happen to be in the way (map 11).

The strongest earthquake in California's recorded history happened in 1857. It was named the Fort Tejon quake, though its epicenter was north of there, near Parkfield, along the infamous San Andreas Fault. Only minor damage occurred, although the 7.8 magnitude quake was felt in Los Angeles, a city of only 4,000 people at the time. The Great Inyo quake was of similar intensity in 1872 at Independence in the Owens Valley.

Another San Andreas Fault earthquake destroyed much of San Francisco. The magnitude 7.7 quake did its initial damage on April 18, 1906, but for days later fires burned and compounded the catastrophe because water pipes that supplied the city's fire hydrants had been destroyed during the shaking (pl. 66). Reservoirs on the peninsula south of the city still held water, but it simply could not be delivered.

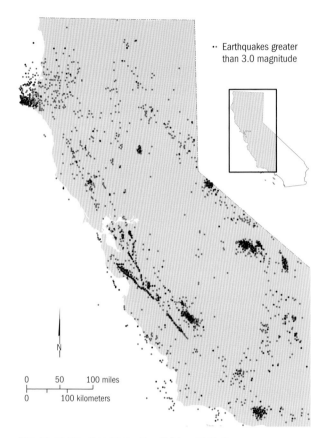

Map 11. Earthquakes greater than 3.0 in magnitude.

The epicenter of that quake was north of the Golden Gate, off the Marin coast. The community of Olema, in Marin County, experienced the most severe land displacement—up to 21 feet—a violent shift that is still visible in the form of a kinked fence line near the headquarters of Point Reyes National Seashore. The north end of the San Andreas Fault leaves the land at Tomales Bay, the long narrow inlet that is another manifestation of the fault.

Plate 66. April 18, 1906, as fires burned in San Francisco following the earthquake. This view was from Golden Gate Park.

Other large earthquakes did extensive damage in San Fernando in 1971 (magnitude 6.7) and in Northridge in 1994 (also magnitude 6.7), both in Los Angeles County.

The 1989 Loma Prieta quake (magnitude 7.0) struck in the early evening as a World Series baseball game was about to start between the two Bay Area teams, the San Francisco Giants and the Oakland A's. At least 60 people died, most of them in the collapse of an elevated section of freeway near Oakland. Also, part of the eastern span of the Bay Bridge collapsed, closing the bridge for more than a month. The Marina District in San Francisco sustained heavy damage where structures had been built on former bay land, filled many decades earlier (pl. 67). Such filled land loses its solidity, undergoing liquefaction and amplified movement during earthquakes. The epicenter was to the south in the Santa Cruz mountains, where the cities of Watsonville and Santa Cruz sustained damage. In total, about 16,000 homes and apartment units were destroyed or damaged. The American Red Cross operated 45 shelters for months that housed, at the peak, 6,000 people.

Plate 67. In the 1989 Loma Prieta earthquake, an automobile was crushed when a three-story apartment building collapsed.

The San Andreas Fault zone marks the boundary between two of Earth's tectonic plates (map 12). The Pacific plate, expanding away from growth points out in the Pacific Ocean, encounters the North American plate near the California coast. Though the North American plate overrides the Pacific plate along much of that boundary, the behavior is different along the San Andreas, which is called a "strike-slip" fault. The Pacific plate moves northwest at about 1.8 inches per year relative to the North American plate. At that rate, in about 12 million years Los Angeles, on the Pacific plate, could end up west of San Francisco, which is on the North American plate. As tectonic plates move against each other, stress builds and faults are said to become "loaded." Earthquakes eventually relieve some or all of that potential-energy load.

As the San Andreas Fault approaches southern California, the line it has followed from the north bends, crosses between the San Gabriel and San Bernardino mountains at Cajon Pass, and then follows more of a southeastern course.

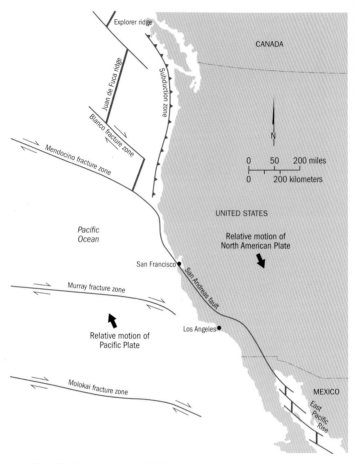

Map 12. Movements along California faults.

Because of that angle change, northwest movement of the Pacific plate generates compression across the fault. The Transverse Ranges are uplifted by that pressure; the Santa Monica mountains rise one inch every thousand years.

The most active creep along the fault occurs in its middle section, south of Parkfield and north of San Juan Bautista.

Plate 68. The San Andreas Fault is clearly visible where it crosses the Carrizo Plain.

Other sections are locked, showing little regular movement since 1906 in the north, and since 1857 for the southern section. The area around the small town of Parkfield, north of the Carrizo Plain, where the San Andreas Fault can clearly be seen from the air (pl. 68), has been studied and monitored extensively by the U.S. Geological Survey (USGS) in an effort to develop earthquake prediction techniques (without much success, to date). Another kink, this one in the highway bridge on the south approach to Parkfield, shows where movement of the Pacific plate bent that end of the formerly straight bridge toward the north (pl. 69).

The San Andreas Fault is the focus of much research and receives a lot of media attention because it is expected to produce "the big one" someday, another major earthquake

Plate 69. The Parkfield bridge was bent by lateral movement of the Pacific plate relative to the North American plate.

approaching magnitude 8 and causing tremendous damage. But the San Andreas is not the only fault in the state and not the only one causing concern. Nearly two million people live on top of the Hayward Fault, along the east side of San Francisco Bay, where major quakes have occurred about every 140 years. A magnitude 7 quake on that fault today could leave 100,000 people homeless and cause $1 trillion in damage, according to the Association of Bay Area Governments and the U.S. Geological Survey (map 13).

The Richter scale is no longer used to describe the size of earthquakes. Today geologists use the "moment magnitude scale," which is considered a better measure of the energy released during an earthquake and more universally adaptable outside California (where Charles Richter developed his scale in the 1930s). Because the scale is logarithmic, each whole-number increase represents 30 times more energy. The new system was designed to be comparable to the more familiar Richter numbers, and the two scales produce virtually the same number.

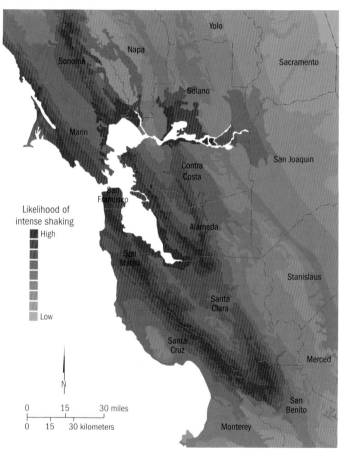

Map 13. Bay Area shaking.

Many news organizations have dropped the use of "Richter" but also avoid use of the phrase "moment magnitude." Instead, they now commonly refer to a "magnitude 7 earthquake," which would release 30 times the energy of a quake of magnitude 6.

The intensity of shaking and the damage caused to structures is related to the magnitude of the earthquake, the distance to the point where the quake originated (where a fault is said to have "ruptured"), and the soil composition beneath a structure. There is much less shaking on hard bedrock than on soft soils, which can actually amplify the movement (as happened in the Marina District of San Francisco during the 1989 Loma Prieta earthquake).

Besides causing structural damage, earthquake events can start fires, break dams, cause hazardous materials to be released, damage power lines and roads, and generate landslides and tsunamis (tidal waves).

During a quake, people should drop, cover, and hold on. The experts' advice is to stay indoors until after the shaking has stopped. Escape attempts put people in danger of being hit by falling objects, inside or outside buildings. The old advice to stand in a doorway, because walls with door frames were built to be stronger than other walls, is probably not valid with today's construction designs. Instead, avoid exterior walls, windows, hanging objects, mirrors, tall furniture, large appliances, and shelves or cabinets filled with heavy objects. If in bed, stay there and cover your head with a pillow to protect it against falling material.

Before a quake, it is wise to anchor anything that might topple, including hot water heaters, bookcases, plates, and figures on shelves. Falling objects cause more injuries during earthquakes than collapsing structures do. Disaster kits (including first aid materials, food, water, and medicine) are a good idea for this and other emergencies, as emergency medical services may be overloaded and communication lines down. Families should make plans together about where to go and how to communicate if separated, keeping in mind that a quake may happen while family members are at work or school and disruptions could last for weeks.

Standard homeowner's and renter's insurance policies do not cover earthquake losses. After the 1994 Northridge quake in the San Fernando Valley, most of the state's insurance companies stopped selling policies covering earthquake damage. In 1995 the California legislature created a reduced-coverage earthquake insurance "mini-policy" that applied to homes but excluded coverage for non-essential items such as swimming pools, patios, and detached structures. In 1996 the California Earthquake Authority (CEA) was created by law. Insurers in California must now offer earthquake coverage, either their own privately funded earthquake insurance or through participation in the CEA funding pool. Eighty-six percent of California homeowners still do not have earthquake insurance.

Volcanoes

California's volcanoes are part of the Pacific Ring of Fire, a zone of vulcanism that wraps around the western, northern, and eastern edges of the Pacific plate. Where the Pacific plate collides with and is overridden by the North American plate (on which most of California sits), the pressure from above turns solid earth into molten lava, which can emerge where volcanic features provide channels to the surface. Mount Shasta and Mount Lassen are the southernmost of the Cascades volcanoes extending from northern California up to British Columbia.

Mount Shasta is an incredible geological feature because it looms over so much of the northern part of the state. Its peak is at 14,162 feet, almost 11,000 feet above the valley below. Shasta began forming almost 600,000 years ago, has erupted at least once every 800 years during the last 10,000 years, and has become more active during the last 4,500 years, with eruptions about 600 years apart. Its last known eruption occurred about 250 years ago (pl. 70).

The last volcanic eruption in California occurred when Mount Lassen awoke and erupted several times between 1914 and 1917. The only other Cascades volcano active since then

Plate 70. Mount Shasta looms above the northern end of the Sacramento Valley.

Plate 71. The 1914 eruption of Mount Lassen was the most recent volcanic event in California.

is Mount St. Helens, which erupted in 1980 in Oregon. Lassen Volcanic National Park contains more than 30 volcanic domes from eruptions dating back 300,000 years (pl. 71).

Another historically active volcanic region with the potential to reawaken is in the eastern Sierra, between Mammoth Lakes and Mono Lake. A massive eruption event occurred there about 760,000 years ago that formed the Long Valley caldera. A caldera is the depression that results when material is blown up and away in an eruption. The Long Valley eruption was large enough to send ash deposits as far east as Nebraska, producing a detectable layer of reddish tuff in the soil across much of the west (map 14).

USGS scientists focused their attention on Long Valley after four magnitude 6 earthquakes were centered there in May 1980. They discovered that the surface of the ground had risen and was continuing to rise, suggesting that magma below was inching upward and causing the quakes. Mammoth Mountain, where a large ski area operates, is a young volcano on the west rim of the Long Valley caldera, built by a series of eruptions between 220,000 and 50,000 years ago. Other eruptions over the last 40,000 years formed a chain of craters that extends 35 miles to Mono Lake. There has been an eruption somewhere along that chain every 250 to 700 years during the last 5,000 years. The most recent eruptive activity was underneath Mono Lake, only about 250 years ago. The bottom of the lake was lifted to the surface to form Paoha Island, and several craters on that island mark where lava reached the surface (pl. 72).

Given the signs of activity in the region, the USGS calculates that the probability of another eruption in any given year is less than 1 percent, comparable to the annual chance of a magnitude 8 earthquake along the San Andreas Fault.

Since 1980, when Mount St. Helens erupted in Oregon and the Long Valley caldera activity was noticed, the USGS has monitored both Lassen and Long Valley to detect signs

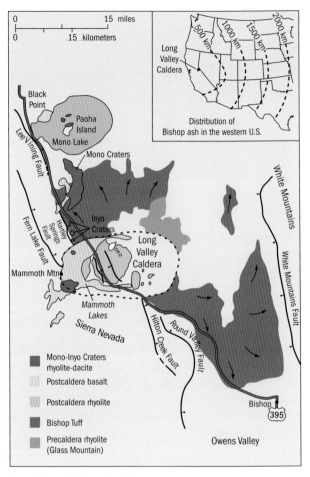

Black Point

Paoha Island

Mono Lake

Lee Vining Fault

Mono Craters

Hartley Springs Fault

Fern Lake Fault

Inyo Craters

Long Valley Caldera

Mammoth Mtn

Mammoth Lakes

Sierra Nevada

Hilton Creek Fault

Round Valley Fault

White Mountains

White Mountains Fault

Bishop
395

Owens Valley

Distribution of Bishop ash in the western U.S.

Long Valley Caldera

500 km   1000 km   1500 km   2000 km

- Mono-Inyo Craters rhyolite-dacite
- Postcaldera basalt
- Postcaldera rhyolite
- Bishop Tuff
- Precaldera rhyolite (Glass Mountain)

Map 14. Long Valley volcanic features.

of new volcanic activity. Periodic field measurements are made for ground deformation and volcanic gas emissions, and local instruments continuously transmit data by satellite to USGS offices in Menlo Park.

Plate 72. Paoha Island and Panum Crater (in the foreground) are young volcanic features at Mono Lake in the Eastern Sierra. The eruptions that formed them occurred only a few hundred years ago.

Landslides

When the earth moves suddenly and powerfully, human structures and lives can be lost. The exceptionally heavy rains that came with the El Niño winter of 1982–1983 caused landslides that damaged or destroyed 6,300 homes and 1,500 businesses across the San Francisco Bay Area. This region is susceptible to landslides because the Coast Ranges hills are carved into steep-walled canyons in many places. After rain saturates the soil, additional heavy rain leads to landslides. In early 1983 more than 4,000 people used American Red Cross evacuation shelters. There were 33 deaths caused by the storms and 25 of these were caused by debris flows, where mud carried trees and boulders down mountain sides. The storms caused $300 million in damage. Ten people died near Love Creek in the Santa Cruz mountains, when a hillside let go to produce a 20-foot-deep wall of debris that carried trees, boulders, and mud.

There are two types of landslides: (1) sudden, rapidly moving debris flows that occur during heavy rain or a few hours after the rain stops; and (2) deep-seated, mass-wasting landslides that move more slowly and may cover much larger areas. The deeper slides present less immediate danger to people because they move more slowly, but their damage can extend across broader areas. In both cases topsoil becomes saturated from successive winter storms, and the next big storm can trigger a release of the saturated soil above underlying bedrock, which does not absorb water as readily.

Another El Niño scenario set up in the eastern Pacific Ocean for the 1997–1998 winter. With such comparatively warm ocean water cycles, northern California typically receives above-average precipitation. On February 2, 1998, a storm triggered both types of landslides, again in the Santa Cruz mountains. Homes at the bottom of draws were inundated by debris flows. Deep-seated landslides also affected parts of San Mateo County. Homeowners first noticed doors sticking, door frames becoming distorted, and cracks forming in their driveways. Some homes twisted and shifted on their foundations. Eight homes became dangerously unstable, and the county ultimately rezoned the area as permanent open space.

In southern California, the small residential community of La Conchita was subdivided in 1924 on a narrow strip of coastline backed by a 600-foot cliff. Historical accounts back to 1865 described landslides periodically coming off that cliff. The Southern Pacific rail line was covered by slides in 1889 and 1909. In 1995, 14 houses were damaged or destroyed in La Conchita by a slow-moving, deep slide. In 2005 a landslide moved much more rapidly and with more catastrophic consequences. It had rained for 15 days; nearby Ventura measured more than 19 inches of rainfall (the average for that period was less than five inches). A landslide raced downhill and overtopped a protective wall that had been constructed after the 1995 event. The landslide

Plate 73. The La Conchita landslide in 2005.

destroyed 36 houses and killed 10 people. The county declared the area a geological hazard area, but residents still live there. Some residents have sued the county for building roads and making public improvements that facilitated building in La Conchita (pl. 73).

Preparing for landslides is a challenge. The USGS has mapped hazard zones (map 15) and advises that culverts and ditches be kept clean so that water does not pond on property. Construction of deflection walls might be considered (with proper attention to where the deflected water will go). Measures to manage water can be very expensive, with no guarantee that they will be effective. The USGS also suggests that, when conditions become threatening, people evacuate. Possessions and structures can be replaced, but lives cannot.

Reality Checks

The nation's taxpayers must wonder sometimes why they shoulder the federal government's disaster relief costs that so often serve Californians primarily or exclusively. Rebuilding

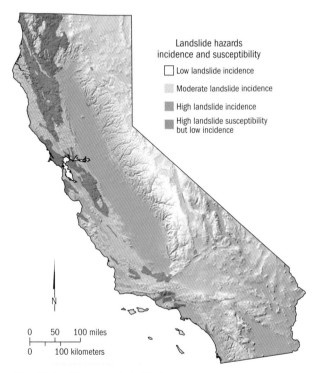

Map 15. Landslide areas in California.

with government funds managed by the Federal Emergency Management Agency (FEMA) is seldom enough to fully restore what people have lost. Disaster relief funds go chiefly toward low-interest loans—that must be repaid—for repairing and replacing property not covered by insurance. Grants can be made to individuals and households who do not qualify for loans. The average FEMA grant is less than $15,000 and the maximum is $26,200. The maximum personal-property loan is currently $40,000 and the maximum for primary home repair is $200,000. In California, those amounts may not come close to allowing full replacement.

Perhaps there should be zoning to control development in areas deemed at risk from natural hazards. That has always been a controversial idea in California, where so much of the economy has been driven by real estate development ever since statehood. Whenever there is a disaster—be it wildfire, earthquake, landslide, or flood—television reporters invariably interview one or more individuals preparing to rebuild on the same piece of land. The untamed land could teach us humility, but our society seems to admire defiant self-reliance. Most of us are content to balance the good life in California against the odds that a disaster will strike once or twice in our lifetime.

Dirt made my lunch,
Dirt made my lunch.
Thank you Dirt, thanks a bunch,
For my salad, my sandwich
My milk and my munch, 'cause
Dirt, you made my lunch.

If one accepts the proposition (and many would not)
that humans are a part of nature, and that farming is a
natural activity of humans, then it follows that for some
parts of the land agriculture is a natural state. Farming
is not a perversion of nature, but a natural development
in our planet's evolution.

MADISON (2002, 51)

FOOD DOES NOT simply appear on store shelves. Agricultural production, our source of food and fiber, requires healthy soils and the lion's share of the state's developed water supply. The state's farm products reach far beyond California; 55 percent of the nation's produce is grown here. Agricultural lands also provide open space that has qualities different from those of wildlands, but also a much different feel compared with urban and suburban development. By giving sufficient attention to soil health, farms can participate in carbon sequestration efforts that help address the global climate change crisis.

Why Farmlands Matter

California agriculture generates about $39 billion a year. It is a unique "industry" because of its biological imperatives—the majority of farm products provide fundamental means of life support. If urban communities are the "backbone" of the California economy, farms can be considered its vital organs.

California has been the leading agricultural state in the nation for over 50 years. It is the number one dairy state, with cows converting plant products grown in the soil into milk, cheese, and yogurt. California farm soils are unusually diverse and that translates into an amazing array of crops. A large number of specialty crops are produced almost exclusively in California, including more than 98 percent of the nation's almonds, artichokes, dates, figs, olives, pistachios, prunes, raisins, and walnuts (pl. 74). Nine of the 10 top agricultural counties in the United States are in California. Fresno County was number one in 2009 (with about $3.4 billion in farm revenues), followed by (in order) Tulare, Kern, Monterey, Merced, Stanislaus, San Joaquin, San Diego, Kings, and Imperial counties.

Plate 74. An orchard in the Sacramento Valley.

Although agricultural land is steadily being lost to urban encroachment, in 2009 the state still had 29 million acres of agricultural land, about one-third of its land area. Croplands occupied 12.5 million acres and livestock grazed on the other 16.5 million acres. There were almost 80,000 farms and ranches, but the great majority of the acreage was divided among about 5,000 particularly large land holdings.

Soil differences account for most of the categories used to differentiate important farmlands. The distinctions become important when one considers just how much is lost when specific agricultural lands are surrendered to urban encroachment. For example, *Prime Farmland*, as identified by the California Department of Conservation, is limited to just 5,274,000 acres in California (out of about 100 million total acres of land in the state). Prime Farmland has the optimal physical and chemical features for sustained high yields and long-term agricultural production, given the soil quality, growing season, and local moisture supply (map 16, pl. 75) *Farmland of Statewide Importance* (2,714,000 acres) is similarly arable, but with minor shortcomings such as steeper slopes or less ability to

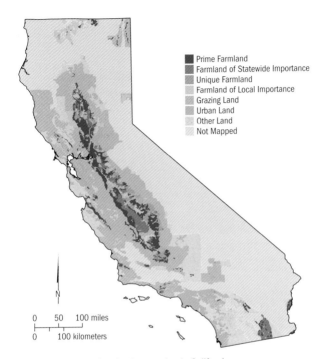

Map 16. Important farmland categories in California.

store soil moisture. The *Unique Farmland* category includes 1,300,000 acres with poorer-quality soil that usually must be irrigated. A considerable amount of such land has been devoted to vineyards in recent decades. These first three categories cover most of the floor of the Central Valley, Salinas Valley, Santa Maria Valley, the Oxnard and Pajaro plains, and the Imperial Valley south of the Salton Sea. The *Urban Lands* of Los Angeles and Orange counties and the Silicon Valley region of Santa Clara County once had similarly valuable agricultural soils. Los Angeles was the leading agricultural county in the nation during the first half of the twentieth century.

Plate 75. Lettuce in the Salinas Valley.

There are also *Farmlands of Local Importance* (2,855,000 acres) serving particular local agricultural economies, a status determined by county boards of supervisors and local advisory committees. *Grazing Lands* outline the valleys, generally within the Coast Ranges and the foothills of the Sierra Nevada.

Soil character, in combination with climate and topography, determines what can be grown and where. In the Sacramento Valley, annual rainfall comes entirely during the winter months. There are only about 2,000 acres of non-irrigated farmland there. Wheat and barley are grown without irrigation by sowing seeds in the fall that are watered by winter rains; harvesting occurs in early summer. Walnut and peach orchards are typically located near Sacramento Valley river channels to take advantage of the well-drained alluvial soils. With irrigation, an incredible variety of crops becomes possible. Growing rice requires seasonal flooding of fields. Rice is one of the lower valley's principal crops, grown on relatively impermeable, waterlogged soils that once were seasonal wetlands. Draining wetlands and confining rivers behind levees to convert floodplains into farmland is the principal reason why, today, California retains only 5 percent of its original wetlands.

In the hot desert land of the Imperial Valley, irrigated farmlands flourish on sediments deposited over centuries by the Colorado River. The warm climate there allows multiple crops to be grown throughout the year, with the leading products being alfalfa, lettuce, and broccoli. Imperial County annually generates $1.37 billion in farm products (pl. 76).

Though there is no organic-rich "topsoil" in the Imperial Valley desert, centuries of river flooding has filled the basin

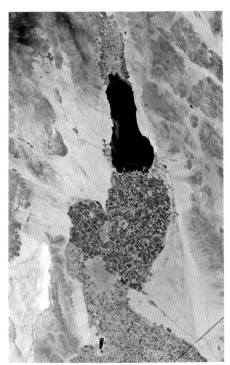

Plate 76. From space, the farms of the Imperial Valley are evident south of the Salton Sea. North of the sea, more farms appear in the Coachella Valley.

with soil deposits more than a mile thick. Whenever the silt-laden Colorado River, on its way to the Gulf of California, built up enough sediment deposits to dam the regular channel, the river would change course and, for a while, flow northwest into the below-sea-level Salton Sink.

In addition to alluvial soil layers, floods also deposited salts. Water in the Colorado River is heavily loaded with eroded minerals. Plant roots separate irrigation water from the dissolved salts it carries, leaving the salts behind in the soil. Unless the salts can be flushed away from the root zone, crops will suffer. Roots in soil that is saltier than the plant tissues cannot absorb water out of the ground and those plants will wilt or die. Salt levels can reach the point where land is ruined for growing crops. Imperial Valley farmers reliant on Colorado River water battle soil salinity by installing sub-surface "tile" drainage systems under their fields and then applying more irrigation water than the plants require so that dissolved salts leach down below the root zone into the drains. This saline water is then sent toward the Salton Sea.

In the San Joaquin Valley—the southern half of the great Central Valley—table grapes, almond and pistachio nuts, and a host of vegetable crops, including tomatoes and peas, are grown (pl. 77). Farms near Fresno produce raisins. Thirsty alfalfa crops support an expanding dairy industry, as urban sprawl and groundwater pollution have forced dairies out of Riverside County in southern California. Cotton was once

Plate 77. Harvesting peas in the Central Valley

grown on over a million acres in the San Joaquin Valley; however, land devoted to that crop has declined to about 100,000 acres in recent years as farmers shifted to higher-value crops, including pomegranates and permanent nut tree orchards.

About 1.5 million acres in the western part of the San Joaquin Valley have soils created by marine sediments that were naturally saline. The problems of high salt content are exacerbated by shallow clay layers below the topsoil that block the percolation of irrigation water. On the east side of the valley, by contrast, granitic soils allow much better drainage and are naturally low in mineral salts. Irrigation of the San Joaquin Valley's drainage-impaired soils leads, as in the Imperial Valley, to the build-up of salts in the soil.

The Westlands Water District, serving the west-side farms of the southern San Joaquin Valley, has some of the most severe salinity problems in California. Farmers there install tile drains below root zones and grow species, such as cotton, alfalfa, and pistachios, that are salt tolerant. When the federal government's Central Valley Project (CVP) was conceived in order to deliver water to the region's farms, the drainage problems were anticipated. The "San Luis Drain" was intended to carry farm field runoff north through the valley to Suisun Bay, where the Sacramento–San Joaquin delta passes water westward toward San Francisco Bay. Eighty-five miles of the drain were built, starting from the south. By 1973 it reached as far as Kesterson, where a national wildlife refuge was established on the wetlands and ponds created by the drain water. Within a decade, grotesque deformities and high mortality were observed in thousands of waterfowl and shorebirds born at the refuge. Bird embryos had protruding brains, missing eyes, twisted bills, and deformed legs and wings. Elevated selenium levels were the cause (pls. 78a and 78b).

Selenium is a trace element found in the valley's west-side soils. Although trace levels of dietary selenium are essential for most organisms, it becomes concentrated to toxic levels as it is passed up through successive levels of the food chain.

Plates 78a and 78b. Black-neck Stilt *(Himantopus mexicanus)* embryos, normal (S-313) and deformed (S-9), from a Kesterson Wildlife Refuge nest, in 1985.

Not only birds were affected; deformities appeared in insects, frogs, snakes, and mammals as well. Concentrated selenium can also harm humans. The drain leading to Kesterson was closed in 1986. Selenium in farm runoff and evaporation ponds across the state has been closely controlled ever since.

A drainage solution had been promised to farmers receiving CVP water, and court decisions confirmed that a solution had to be provided. Sending the salty, toxic drainage to San Francisco Bay is now considered an unlikely option. The Bureau of Reclamation in March 2007 adopted an on-farm disposal plan that would retire 194,000 acres of farmland and create 1,900 acres of evaporation ponds to handle treated drain water, from which salts and selenium had been removed.

Later that year the Westlands Water District suggested a rather bold settlement alternative, proposing to take over financial responsibility for the cleanup effort from the CVP and asking, in return, forgiveness of the $100 million the district still owed for CVP construction, and also that it be granted perpetual federal water rights. The district proposed fallowing only 100,000 of the acres prone to selenium runoff. A decision was still pending in late 2008, when the Central Valley Regional Water Quality Control Board required the Westlands Water District to promptly file for a waste discharge permit and present a plan to finally clean up the illegally contaminated soil and water.

Dirt First!

> *The underlying problem is confoundingly simple: agricultural methods that lose soil faster than it is replaced destroy societies.*
>
> MONTGOMERY (2007, 141–142)

Traditionally, farmers have prepared the ground before planting by plowing and disking to turn over and loosen the topsoil to uproot weeds, to incorporate crop residues and manure amendments into the soil, and to break clods into finer pieces (pl. 79). Family gardeners have relied on shovels and rototillers

Plate 79. A large farm tractor drags chisels that rip deep into the ground.

to similarly prepare garden beds. Such disturbance of the upper layers of soil, however, can also interfere with life processes that give soil its natural fertility.

Tillage can so reduce the level of organic matter in farm soils that their biological systems are essentially lost. Uncovering and exposing upper soil layers brings sub-surface soil organisms up to where they may dry out or be oxidized. Mechanical disturbance can tear up mycorrhizal fungus mats and kill significant numbers of earthworms. Plowing and disking breaks apart soil aggregates, and this can increase wind and water erosion. Plows turn crop residues under to promote their incorporation into the soil, but moldboard plows can bury residues (and the former topsoil) as much as 14 inches deep. Oxygen levels at those depths are so low that decomposers that normally dwell near the surface suffocate and residue decomposition actually slows.

Farmers face an inherent challenge with regard to maintaining organic matter in soil, because they harvest and remove their crop products, leaving only residues of stems and stubble, seeds and pods, to work back into the ground. In natural ecosystems, by contrast, most plant matter dies in place so that organic matter constantly recycles through the soil.

To increase and maintain organic matter levels, "conservation tillage" techniques involve the use of shallow tillage or no tillage at all. Farmers may till only small strips of land to create a seedbed or drill seeds directly into the untilled soil. Shallow tillage turns residue under to speed decomposition by bringing it into contact with soil microbes and also adds oxygen to that upper layer, which benefits microbes. A cover crop may be grown during the off-season and the commercial crop later planted within the residue from those cover plants.

Soil managed without any tillage relies on soil organisms, such as undisturbed earthworm populations, to incorporate plant residue into the soil. It keeps organic matter and associated nutrients close to the surface, where they will benefit the next round of growing plants. Reduced tillage also minimizes soil compaction, because heavy equipment needs to make fewer passes across the field. Less compaction increases the soil's ability to absorb water quickly, reducing water loss through evaporation and runoff. Less equipment operation means savings in fuel costs and salary costs for equipment operators, and a reduction in the dust that contributes to particulate air pollution, a major concern in the San Joaquin Valley. According to the Natural Resources Conservation Service, farmers can save at least 3.5 gallons of fuel per acre by going from conventional tillage methods to no-till. On one thousand acres of cropland, that translates into savings of 3,500 gallons of diesel fuel per year and many thousands of dollars.

There are other environmental benefits. Less soil disturbance through conservation tillage practices can mean less dust. Maintaining adequate levels of organic matter in the soil also helps prevent erosion and can reduce greenhouse gases by sending less carbon dioxide from diesel fuel burning into the atmosphere and by storing more carbon within the soil.

After many years in crop production, most California soils are far below their carrying capacities for carbon. That means the soils have the potential for greatly increased carbon storage. Conservation tillage techniques can, therefore, qualify for carbon credit trading. Other ways in which agriculturalists can aid in carbon sequestration include converting from crops that are harvested to grassland, which keeps more carbon in place; practicing sustainable management of rangelands; and planting trees on previously non-forested land. Livestock and dairy operations can limit the release of the greenhouse gas methane by using manure digester systems, which can also earn them carbon credits.

Carbon dioxide gas is lost from tilled soil in direct proportion to the volume of soil disturbed. Deep plowing into the subsoil releases more carbon than shallow tillage. No-till practices cause the least carbon dioxide loss and can even increase sequestration of carbon. Growing off-season cover crops and applying manure also raises soil carbon levels.

Increasing the amount of carbon in the soil is a "win-win" scenario, because it not only helps address the global climate crisis, but also increases soil fertility, which improves farm productivity. Weed management may be the biggest obstacle to expanding the practice of no-till farming, and increasing the use of herbicides to control weeds is not an attractive option.

There is great potential for expansion of conservation tillage in California. Although the techniques are widely used in the Midwest, conservation tillage was applied to less than 0.5 percent of row crop acreage in California in 2005. Many vegetable types are grown as "row crops" in this state, and new techniques for conservation tillage with such crops have not yet been developed or tested. Farmers are focused on overall productivity and profit margins, and need to be convinced that the switch to alternative methods is cost-effective and worth the effort.

Conservation tillage presents several challenges. Crop residues left on the ground can slow the flow of water in furrows, requiring a change in irrigation method. Drip irrigation strips are one alternative. Controlling weeds, traditionally done through cultivation, may become a problem that leads to increased herbicide use. Avoidance of deep tillage can increase problems with gophers and other rodents by leaving more attractive organic matter in fields and leaving underground tunnels undisturbed. With a surface mulch in place, the ground may not absorb as much sunlight, so that colder temperatures at the surface may lead to reduced early-season growth and increased risk of frost damage.

Still, some California innovators are developing techniques specific to local requirements. The *California Tillage* newsletter, Winter 2008 issue, reported on the San Juan Ranching Company in Dos Palos, which plants a variety of wheat as a cover crop between regular crop harvests (of tomatoes, cotton, alfalfa, grain corn, chili peppers, and

Plate 80. The peat soil of the Sacramento–San Joaquin delta islands is formed as tules and cattails decompose in soil saturated with water.

melons), taking advantage of a five-foot taproot developed by the wheat to naturally break up the soil.

An interesting application of carbon capture "farming" has begun on islands in the Sacramento–San Joaquin delta. Soils in the delta are peat, the product of decomposition of marsh vegetation such as tules and cattails in wet, oxygen-free conditions (pl. 80). After farming began there, the peat soils became exposed to the air, especially when land sat fallow after harvests. Peat oxidizes on contact with air, a slow "burning" reaction that sends carbon dioxide into the atmosphere. The lost carbon translates into lost soil volume, so the surfaces of delta islands have been subsiding. Today, most Sacramento–San Joaquin delta islands are 10 to 30 feet below sea level and must be protected with surrounding levees and the constant operation of groundwater pumps.

Twitchell Island, in the western delta, is 12 feet below sea level. There U.S. Geological Survey and Department of Water Resources scientists replaced 15 acres of former cropland with ponds nurturing native tules and cattails, an experiment begun in 1997. The land had been subsiding more than an inch each year, but new tule/cattail muck accumulation reversed this process, increasing soil depth more than 10 inches by 2005. Building up the island was a positive outcome; but in addition, the growth and decomposition changed the island soil from a carbon dioxide emission source into a carbon sink, storing carbon extracted from the atmosphere by the photosynthesizing plants. With these positive results documented, plans have been made to expand the effort to 400 acres of nearby Sherman Island.

Preventing Erosion

Bare ground is land at risk. Muddy runoff water and gullies are signs that erosion and topsoil loss are occurring. The U.S. Department of Agriculture has estimated that it takes 500 years to produce an inch of topsoil. Erosion can undo that slow accumulation in an exceedingly short time. Having learned from the mistakes that led to the Dust Bowl–era wind and water erosion crisis, farmers counter topsoil loss by terracing the land and diverting runoff water. But many such standard erosion control practices can be avoided if the ground stays covered year-round and not left bare after crops are harvested. Modern no-till practices, mulching with compost and manure, leaving crop residue in place on the surface, and planting interim cover crops are options available to farmers to prevent soil loss (pl. 81).

Manufactured Fertilizers

To counter the lost fertility that comes with harvesting crops and land-tilling, farmers have relied on regular applications of human-made fertilizers to maintain field productivity. Excessive application of fertilizers can, ironically, further

Plate 81. A cover crop of clover grows between rows of trees in an orchard.

reduce natural soil fertility by introducing salts that kill microorganisms. Fertilizers can also provide an excess of nitrogen that stimulates populations of decomposition bacteria to the point that too little organic matter is available to feed soil bacteria, which then starve. Synthesized fertilizers are salt based. Salt build-up not only destroys a plant's ability to take up water, but can also directly kill mycorrhizal fungi and worms. Nitrogen applications commonly exceed the amounts that plants and soil organisms can process, and the excess dissolves in ground and surface water. Nitrogen leaching into lakes can lead to unnatural algal blooms that can consume all of the oxygen, killing fish and plant life. When nitrogen contaminates domestic water wells, as has happened in communities close to dairies and feedlots with inadequately controlled animal wastes, the "blue baby syndrome" can result. Nitrates interfere with the oxygen-carrying ability of red blood cells and the effect can be especially serious for infants. Most adults have an enzyme that converts the damaged red blood cells back to normal.

Composting takes advantage of soil ecosystem processes to recycle organic "waste" and produce humus for home garden and yard soils, or for community-size, centralized composting operations and large-scale farmland composting. At the individual home level, kitchen wastes need no longer go down the garbage disposal and into the sewage system. Yard clippings need not be added to haul-away garbage. In composting, approximately equal proportions of "green" items (vegetable peelings, weeds, and grass clippings) and "brown" items (dead leaves, wood chips, sawdust, and straw) make an ideal blend that decomposes well and produces a nutritious humus as the end product. Compost material can simply be piled or may be confined within bins. It should be kept reasonably moist and may need to have water added, depending on local climate conditions, but not kept so wet that air cannot move through the pile.

As in natural soil ecosystems, both bacteria and fungi accomplish the breakdown of materials. Fungi work particularly on the cellulose and wood in compost. Earthworms may be introduced to help manufacture humus. Heat is generated when the pile is decomposing properly and thermophilic bacteria are doing their thing. Temperatures can exceed 140 degrees Fahrenheit. Piles should be turned occasionally with a pitchfork or spade to aerate the material and prevent smelly anaerobic decomposition (pl. 82).

Composting solves disposal problems, and the final product fertilizes and mulches planting beds. A "halfway there" bucket with a lid can be kept in the kitchen as a convenient place to scrape food from plates and the cutting board. When it gets full, the contents can be brought outside to the continuously operating pile. A large volume of material will eventually produce a smaller volume of

Plate 82. Turning a compost pile aerates the mixture.

clean-smelling humus, which can be harvested by sifting the pile through wire mesh. Some food wastes should not be put into compost bins or piles. Animal parts, especially fatty ones, and ashes from charcoal briquettes (which are generally coated with fat drippings) should be avoided (although fireplace or wood stove ashes are fine). Synthetic plastics are not biodegradable and, of course, have no place in a compost pile. A styrofoam cup will not turn into humus and, though it may break into smaller pieces, essentially never decomposes. So-called biodegradable plastic has starch added to make it disintegrate; unfortunately, once the starch decomposes, very tiny nondegradable pieces of plastic remain.

Thirsty Farmlands

Raising crops and livestock requires a lot of water. Farmers in California use about eight gallons of water to grow one tomato and over 600 gallons for a burger patty, according to the California Farm Bureau. The food we each consume every day represents about 4,500 gallons of water. About 80 percent of California's developed water supply, that portion of the annual runoff that is captured and transmitted by water delivery systems, goes to agriculture (pl. 83).

California farmers face an ongoing challenge to apply water as wisely as possible to their fields and orchards. With so much water going to farms, the burgeoning cities and sprawling suburban regions of the state keep a constant eye on agricultural

Plate 83. Water is siphoned from a canal to irrigate vegetable crops.

practices and agricultural water rights, to see what might be shifted their way. Each time one of the state's cyclical droughts hits and water supplies tighten, some northern California farmers fallow land or shift from the use of surface water to groundwater pumping so that they can sell water to cities, transporting it there via the state's aqueducts.

Statewide, groundwater overdrafting is a problem when pumping exceeds the local groundwater recharge rate. There are no underground "lakes" or streams; groundwater basins or aquifers are simply areas where soil and sub-soil pores are saturated with water that can be tapped by wells. Over-drafting has caused ground subsidence at a shocking rate. In the San Joaquin Valley, near Mendota, land subsided nearly 30 feet between 1925 and 1977 (pl. 84). Overdrafting in the Santa Clara Valley in the 1930s caused the land around San Jose to subside so much that high tides in San Francisco Bay became a threat to the city.

Plate 84. Near Mendota in the San Joaquin Valley, the land subsided nearly 30 feet because of groundwater pumping between 1925 and 1977.

During droughts, groundwater pumping increases and the overdrafting problem is exacerbated. From 2003 to 2009, the Central Valley groundwater basin lost 24 million acre-feet of water, as estimated by space satellites that sense changes in gravity as groundwater is pumped away. That rate of loss is equivalent to California's entire annual allotment of Colorado River water. Fortunately, depleted groundwater basins can generally be recharged. One of California's greatest opportunities for "new" water storage is to take advantage of wet years, when surface supplies are adequate, to store water underground in the depleted aquifers.

The competition for water rights is one of the greatest forces behind the conversion of farmlands to serve urban growth.

Death by a Thousand Cuts

> High quality farmland is being disproportionately selected for development . . . largely because most California cities are located . . . where our agrarian ancestors settled precisely because of the fecundity of the land.
>
> AMERICAN FARMLAND TRUST (2007, 8)

> To many of us, the rural landscape has an aesthetic and cultural value equal, in a subtle way, to the national parks, and we see it as being equally worthy of preservation.
>
> MADISON (2002, 22)

Neither the United States government nor the state government of California has a policy for designating valuable farmland as too precious for conversion to other uses. The most fertile soils, those of the great Central Valley, could theoretically be declared an essential public treasure and farmland there given permanent protection from urbanization pressures. Instead, 60 percent of the development in the San Joaquin Valley between 1990 and 2004 took place on the highest-quality farmland. Over a half million acres of California's farmland were lost to urbanization during that period. The American Farmland Trust (AFT) considers the Sacramento and San Joaquin valleys to be the nation's most threatened farm region, not only on the basis of acreage lost, but also because the Central Valley is the nation's most important agricultural resource. An AFT report in 2007, *Paving Paradise: A New Perspective on California Farmland Conversion*, predicted that by 2040 the Central Valley could lose another one million acres, or 15 percent of its remaining farmland. The population of the Central Valley could triple by then if the most worrisome estimates prove true—if we choose to let that happen.

The region around Fresno, in the center of the San Joaquin Valley, illustrates the trend. Since the state's Farmland Mapping and Monitoring Program began in 1984, Fresno County has led California counties in urbanization of irrigated farmland. One of the fastest-growing regions in the nation over the last two decades, Fresno County is also the nation's leading agricultural county, producing over 250 types of crops totaling $3.4 billion in value each year.

Another hotspot in the valley surrounds the city of Tracy in San Joaquin County, where affordable housing has attracted residents willing to commute for hours each day to jobs in the Bay Area. Throughout the county, urbanization swallowed 16,000 agricultural acres between 1990 and 2002. Prime Farmland acreage actually declined by over 22,000 acres, because some farms stopped active production (pl. 85).

According to the AFT report, one-sixth of the land development in California since the Gold Rush occurred between 1990 and 2004. The average rate of farmland loss in modern times is 38,448 acres per year. Statewide, development is consuming an acre of land for every 9.4 additional people. At the current rate, another two million acres of California land could be paved over by 2050.

Plate 85. Subdivisions in western Tracy where the land had been farmed.

There is still a lot of good farmland in the state. Some people, even some in agriculture, see no reason to worry about the lost acreage. Even so, the fully urbanized southern California basins and the Santa Clara Valley—now better known as "Silicon Valley," south of San Francisco Bay—once were leading agricultural production areas for the nation (pl. 86), and transformations such as these should serve as a cautionary lesson about complacency.

Though most of the change in southern California followed World War II, 11,500 acres of farmland in Orange County were lost between 1984 and 2002—open lands that had become precious relief from the never-ending onslaught of "progress." With the inclusion of development on land that had not been used in farming, the county urbanized 33,500 additional acres during that 18-year period.

Complacency toward these losses also neglects the fact that development planning can lock in an agenda for growth that extends to decades. Preservation planning needs to be just as farsighted.

Plate 86. Oil wells in an orange grove in 1932, Atwood, Orange County.

Preservation Planning

A measure to control farmland loss was approved by voters of Ventura County in 1998, who stripped their Board of Supervisors of the power to approve new subdivisions on land zoned for open space or agriculture. With the passage of the SOAR (Save Open-Space and Agriculture Resources) initiative, proposals for rezoning must receive voter approval. In the same election, four cities in Ventura County also voted to confine their future growth within specified boundaries.

Public and privately funded land trusts have protected thousands of agricultural acres in Sonoma, Yolo, Monterey, Marin, and other counties by establishing conservation easements and paying farmers to agree never to sell to developers. The Marin Agricultural Land Trust, formed in 1980, was the first of its kind in the nation. It has preserved over 40,000 acres, more than 40 percent of the agricultural land in Marin County, through permanent easements that stay with the land parcel as ownership changes.

The University of California at Davis has a policy of protecting an equal amount of prime farmland for every acre consumed by campus expansion, and the city of Davis ups that to two acres of farmland for each acre developed. Davis and the nearby city of Dixon have established an open-space and agricultural buffer between them utilizing farm easement agreements.

Sixteen million acres of California farmland are covered by the historic 1965 California Land Conservation Act, also known as the Williamson Act, which provided property tax breaks for farmland owners who voluntarily agreed not to sell to developers for 10-year blocks of time. The land was also protected from annexation by cities or special districts. Taxes on the land were based on agricultural income rather than open-market values. The annual tax savings for farmers and ranchers amounted to as much as 83 percent, according to the California Department of Conservation, which administers the program.

Because 10 years is actually a very short period of preservation in the context of urban growth decision making, the alternative "Super Williamson Act" was made law in 1996. It provided heftier incentives for those willing to lock away their development rights for 20 years in Farmland Security Zones. It was also a voluntary program: willing landowners received a one-time payment in exchange for giving up the right to convert their land to non-agricultural uses. In both programs, another year was added to the protection period annually, unless the landowner announced the intention to end the contract at the end of the next cycle. The conservation contracts continued if the land was sold to another farmer. "Permanent" protection in the latter case allowed a review of the agreement after 25 years.

Both of these farmland conservation efforts slowed, but obviously did not stop, agricultural conversion (pl. 87). Almost half of the state's agricultural land has not been enrolled in either program. Moreover, as this book was

Plate 87. Houses encroaching on farmland.

being prepared, in response to the California budget crisis the state stopped funding Williamson Act contracts. Counties will no longer be reimbursed by the state for revenues lost from farmers in the program paying less in property taxes. When the 10-year contracts expire, if the program is not revived, the pace of farmland conversion could rapidly accelerate.

The End of Agriculture?

Such an outcome may actually please those economists who have argued that farming does not represent the highest use of California's land, proclaiming that the relentless conversion of farmland to urban growth is a measure of "inevitable progress." Let other states or nations grow our food and fiber, they suggest, while we stick to our areas of competitive advantage: producing innovations like those that have emerged from Silicon Valley. "We must learn to let go of farming and ranching," Steven Blank wrote in *The End of*

Agriculture in the American Portfolio. "In the short run, this means eliminating the subsidies that delay the inevitable development of our nation out of agriculture and into more profitable industries. In the long run, this means becoming citizens of the world, dependent on others for our food commodities while we produce the marvels and know-how for the future. We have to do these things to become King of the Hill. The first step is to accept that farming, although it enabled us to move into our dynamic future, is part of our proud past" (Blank, 1998, 195–196).

That sounds like biological suicide, given the world's vital need for food and fiber. What about the insecurity inherent in relying on other nations? What about the benefits of agricultural open space as relief from unbroken urbanization? And what about the living soil that would be lost and the effects of this on the extended ecosystems supported by that earth?

Memo to economists from the real world of ecology: Factories do not carry on photosynthesis or restore oxygen to the atmosphere. We cannot eat our computers.

WALKING SOFTLY
California's Footprint on the Earth

*Do we not already sing our love for and obligation to
the land of the free and the home of the brave? Yes, but
just what and whom do we love? Certainly not the soil,
which we are sending helter-skelter downriver. Certainly
not the waters, which we assume have no function except
to turn turbines, float barges, and carry off sewage.
Certainly not the plants, of which we exterminate whole
communities without batting an eye. Certainly not the
animals, of which we have already extirpated many of
the largest and most beautiful species.*

In short, a land ethic changes the role of Homo sapi-
ens *from conqueror of the land-community to plain
member and citizen of it. It implies respect for his fellow-
members, and also respect for the community as such.*

LEOPOLD (1966 [1949], 239–240)

Mount Trashmores

Californians today generate 93 million tons of waste each year, averaging six pounds per person each day. Waste disposal historically has involved shoving garbage "over the side," into handy ravines that became "dumps." Burnable wastes were incinerated, a practice that had to be restricted in California after World War II because of air quality concerns. Back in 1873, when Los Angeles had only 6,000 residents, the city's first garbage and dead animal plot was established. Up in San Francisco, wastes (along with scuttled ships!) became part of the landfill content along the waterfront, pushing land further out into the bay, especially at the foot of present-day Market Street and the Marina district on the north side of the city. Diking and filling all around San Francisco Bay gradually reduced the open-water area from 787 square miles to 548 square miles. Before 1965, bay shore filling averaged 2,300 acres each year. That year legislation was passed to create the San Francisco Bay Conservation and Development Commission, with a mandate to control bay fill activities.

Today most of the state's trash is trucked to tightly regulated regional landfills. When a landfill has taken all the trash it can handle, it is capped with a final layer of dirt, revegetated, and then monitored for leaching of liquids, gases, and slope stability. Such sites have become parks, golf courses, or non-irrigated open space—permanent monuments to modern waste management practices. An example is the hill called "Mount Trashmore" by the citizens of Alameda in the East Bay, now open to hikers, joggers, kite fliers, and dog walkers (pl. 88).

Landfills are required to be lined with clay and plastic to prevent leaching into groundwater. Each day's deposits are covered with earth to keep material from blowing, sliding, or being carried away by animals and to facilitate decomposition. In landfills decomposition occurs through the same processes as elsewhere in the soil. Biodegradable wastes are worked on

Plate 88. Mount Trashmore, where residents of Alameda once took their trash, is now fenced off and vented, the trash covered with soil.

by bacteria and fungi. Microbes break down paper, wood, and plant wastes and then others convert sugars into acids. Methanogen bacteria convert the acids into methane gas and carbon dioxide. At some landfills, methane is captured and burned to generate power. Otherwise it simply escapes and adds to greenhouse gas levels in the atmosphere.

Although decomposition in landfills relies on the living soil's usual cast of characters, landfills have particular issues associated with toxic materials. Many items are banned from the trash, including batteries, fluorescent bulbs, and electronic equipment, such as computers and televisions, that contain heavy metals. The California Integrated Waste Management Board (CIWMB) developed the Take It Back program to identify business partners that will accept "e-waste" and pass it along to appropriate locations for processing.

Waste management is "integrated" when it includes recycling paths to reduce the volume of waste material that goes to landfills. As of 2008 California was diverting more than

58 percent of its waste to recycling and led the nation in both recycling and reduction of landfill deposits. After garbage is picked up, it may be sorted in a materials recovery facility, where recyclable material is separated out and the rest sent on to a landfill. The CIWMB's efforts are funded through recycling and dumping fees rather than through the state's general fund; this was an agency on the cutting block in the state's grim 2009 budget negotiations.

To control the volume of wastes generated by the 38 million people in this state, individuals can "reduce, reuse, and recycle." The most obvious step is to take advantage of the community recycling program, sorting material into appropriate bins, depending on local requirements. Composting yard and food wastes keeps them out of the garbage (and will ultimately benefit landscape plants and gardens). Leaves and grass clippings that are not composted can be spread as mulch. As consumers, we can buy items made from recycled materials and seek products that come with minimal packaging. The convenience of "disposable" products is a problem, when utilized by millions of users. Durable, reusable containers and jars cut down on the use of non-degradable plastics. Special sites are provided in each community to accept discarded oil and old tires, which should not go into landfills. Measures such as these will reduce our individual "ecological footprints" on the Earth.

Ecological Footprints

In national parks we are told, "Take only photographs, leave only footprints," but footprints are not without impact. Recent research has shown that the effects of a foot trail across a meadow in the Sierra Nevada extend far beyond the trail itself; insect diversity diminishes for many yards on either side as a result of people regularly walking on a beaten path (pl. 89).

The concept of an ecological footprint uses a complex calculation to attempt to answer a simple question: how much

Plate 89. Footprints impact trails, and the land nearby is also altered.

of the Earth's resources does it take to support our lifestyles? Everybody has an impact on resources because we each consume food, fiber, water, and minerals, and also benefit from the services of the living soil, including decomposition and the cycling of nutrients and gases between the soil and the atmosphere. The ecological impact of an individual, city, or country corresponds to the ecological processes and natural resources required to keep each one going, to produce everything consumed, and to absorb the wastes generated. On a global level, ecological footprint calculations estimate biocapacity, or supply, for cropland, pastureland, forest, fisheries, and carbon storage areas. Supply can then be compared to the footprint, or demand for resources.

The calculations convert all the Earth's resources into a single unit of measurement: productive hectares of land (one hectare equals 2.47 acres). Each person's share of the total productive or "global hectares" is the person's ecological footprint. With about 6.8 billion people on the planet, there are approximately 1.8 global hectares per person available. Yet the average U.S. citizen's ecological footprint is 9.6 global hectares. If everyone on Earth lived like we do in this nation, we would need four more similar planets!

The value of this calculation is the recognition of the choices that individuals can make to minimize their ecological footprints. A number of organizations offer online calculator "exams," some that collapse the number of questions to greatly simplify the process and others that are much more thorough. At www.myfootprint.org, 27 detailed questions are given. Studying the options presented for each multiple-choice question provides a lesson in alternative behaviors and impacts on the environment. The personal footprint calculation at the end of the process is compared against the world average and, for most of us who live in the United States, the sobering revelation is summarized in terms of the "number of Earths it would take if everyone on the planet lived my lifestyle" (fig. 10).

Questions address categories that deal with food (pl. 90), transportation (including carbon emissions), housing (including energy use), and goods and services (what it takes to manufacture and transport items and to deal with wastes). A personal footprint calculation is summarized in global hectares and related to Earth's total capacity. Proportions of that

Figure 10. Ecological footprint calculation results for the average Californian.

Plate 90. A farmers market in San Francisco offers the chance to buy produce grown locally, reducing the carbon footprint associated with food.

footprint that relate to cropland, pastureland, marine fisheries, and forestland are also shown. Similar analysis has been developed to apply to businesses, communities, governments, and schools.

Footprint calculations that focus specifically on tons of carbon emissions per year have been developed recently to help individuals understand their options in the context of the global climate crisis. A carbon footprint calculator is available that is specific to California households and businesses (at www.coolcalifornia.org). Partners in the development of that footprint calculator include the California Air Resources Board, the California Energy Commission, the Berkeley Institute of the Environment, Lawrence Berkeley National Lab, and Next 10 (a nonpartisan organization focused on improving the state's future).

California households average 38 tons of carbon emissions per year; the national average is 42 tons, whereas the world average is 8 tons per year. Results at this site are broken down into the categories of transportation, housing, food, goods, and services. Comparisons with other households

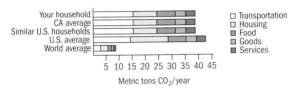

Figure 11. Carbon footprint results: California, United States, and world averages.

in the state and the nation give interesting benchmarks, but again, comparison with the (far lower) world average is bound to be sobering for anyone in California (fig. 11).

An important element of all of the ecological footprint calculation websites is the question "What next?" What can be done with the information? Obviously we do not have multiple Earths to fill the resource needs of humanity. Helpful suggestions are provided for reducing our ecological footprints and moving us toward sustainable lifestyles, but there is something missing from most of these suggestion lists. Individual ecological footprints translate into regional, national, and global crises because there are so many human feet. Population size is a sensitive topic that receives too little attention.

Compassionate Numbers

Everyone now living who was born in 1965 or before has seen the human population more than double, from 3.3 billion in 1965 to 6.8 billion as of 2009. When California became the most populous state in the nation, in 1962, Governor Edmund G. Brown called for a three-day "California First Days" celebration. After hearing about Brown's proclamation, former governor Earl Warren said, in a speech the next day, that he thought the governor was mistaken in his assessment of the importance of mere growth, that there was no merit in simply being the largest. Warren later recalled how he "told them that instead of dancing in the streets, we should ... call the people of California to the schools, churches, city halls, and other places of public assemblage, there to pray for the vision and the guidance to make California the finest state in the Union as well as the largest" (Warren, 1977, 227).

Unfortunately, that kind of visionary, long-term planning never took place. The state's population, not quite 20 million in the 1970 census, had grown to 38.1 million by December 2008, according to the California Department of Finance (map 17). It is projected to grow to 49 million in the next 20 years and perhaps reach 60 million by 2050. Such projections basically extend current growth rates out as if they were a certainty.

Traveling across California, one may observe the amount of open space still left and conclude that there is plenty of land to accommodate population growth. That visual perception ignores the fact that soils vary in their ability to support life. Besides topography, climate, and farmland issues, one of the most important considerations is that much of the land in California has too little water to support a large community. Human ingenuity overcame such resource limits in southern California only after a century of Herculean efforts that included building dams and aqueducts spanning hundreds of miles to bring water into that thirsty region. Without imported water, relying solely on local surface and groundwater supplies, southern California, where about 18 million people now reside between Ventura and San Diego, could support no more than about three million people. The population of California can continue to pursue the water-for-growth paradigm, but only at the expense of agriculture and the environment, where water must be used in other ways.

Some claim that seawater desalination could provide an unlimited source of water for population growth, if energy requirements and costs could be met and the impact on the coastal environment addressed. But is unfettered population growth the future to choose? If it is not, once the benefits of a stable population capture society's attention, embracing

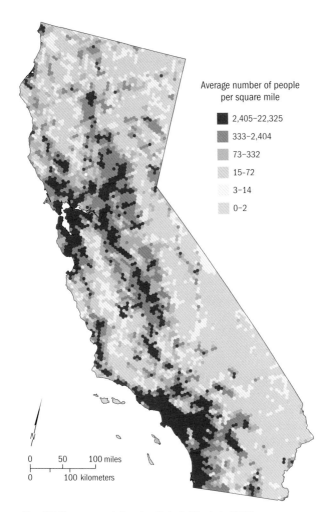

Average number of people
per square mile

- 2,405–22,325
- 333–2,404
- 73–332
- 15–72
- 3–14
- 0–2

0   50   100 miles
0   100 kilometers

Map 17. Human population density in California in 2002.

resource limits could become the means for achieving stabilization without intervention into personal reproductive rights or the institution of draconian immigration restrictions.

Half of Californians (52 percent) in a 2009 survey said that the statewide population growth that has been forecast would be bad for them and their families; 13 percent called it a good thing and 30 percent responded that it made no difference. A two-child family was the ideal size, according to 42 percent of those surveyed. Californians also strongly favored pro-choice policies (Baldassare et al., 2009).

Education for women and access to contraception have been the keys to a phenomenon occurring in at least 51 developed nations where fertility rates are now at or below replacement levels—where populations have stabilized or are in decline. The countries of western Europe, Russia, and Japan fit this trend. Around the world, women bear nearly twice as many children (2.8 per woman) in poorer countries as in wealthier countries (1.6 children per woman).

Fertility rates are currently higher in California (2.1 children per couple) than in the rest of the nation and also higher than in any developed country in the world. Though immigrants new to the state give birth at higher rates, subsequent generations show dramatic declines that conform with the pattern for other states and developed nations.

The state of California is not separate from the world. Our future will be shaped, in part, by the reality that population pressures emerge from regions of poverty. In 2009 a food crisis and epidemics of disease were hitting various parts of the world. Given that California's ecological footprint is already beyond sustainability and that so much environmental damage has occurred here in the most populous state in the nation, a visionary model to guide the state's future seems like an idea whose time has come. Earl Warren saw the need for such guidance nearly a half century ago. Thoughtful, informed choices could take the state along the most compassionate path toward a level of immigration that, in correspondence with the numbers leaving the state, would be consistent with a stable population.

This will be a challenging transition because California's last 150 years of history have been so closely tied to real estate development. In 2002, according to the Department of Finance, one in 44 Californians were employed in the state's $72.5 billion construction industry. That did not include members of the finance, insurance, and real estate industries whose areas of business also relate to development.

Will such a transition ever be initiated? Today the state of California has about 217 people per square mile, mostly concentrated in two "mega-regions." Southern California's mega-region extends from the southern border up through San Luis Obispo and Bakersfield, and sprawls into the Mojave Desert along Interstate 15 (pl. 91). In northern California, the Bay Area is connected to Sacramento and sprawls along corridors of development southward to Fresno in the San Joaquin Valley and eastward on Interstate 80 through the Sierra Nevada foothills. Orange County now has 3,600 people

Plate 91. Sprawling development buries most of the living soil beneath asphalt and concrete.

per square mile, a density greater than in Los Angeles County. The fastest-growing regions in the state recently have been the Sierra Nevada foothills and "the inland empire" areas of Riverside and San Bernardino counties.

Sprawling development not only impacts the land where it occurs, but leads to long, costly commutes that translate into more energy consumption and greenhouse gas emissions. Since the California Global Warming Solutions Act became law in 2006, the state has been committed to reducing its carbon emissions to 1990 levels by the year 2020. Because land use decisions have carbon emissions consequences, Senate Bill 375 was passed in 2008, requiring the state's metropolitan planning entities to demonstrate that future development projects will actually lead to reduced carbon emissions in those regions. Better public transportation systems and homes built closer to work sites are measures that could replace sprawl with what has been called "smart growth." Whether smart or dumb, overall growth still must be factored into our society's ecological footprint calculation.

From Earthworms to Earth Day

For the moment, the phrase "homeland security" verges on the purely propagandistic: who speaks with heartfelt conviction for the health, security, and durability of the American land itself—the soils, waters, plants, animals, and people together? Land, far from being appreciated as the fundamental source of life, health, wealth, stories, poetry, identity, belief, and, yes, patriotism, is just a raw resource and backdrop for busy modern lives.

MEINE (2004, 245–246)

Once in his life a man ought to concentrate his mind upon the remembered earth, I believe. He ought to give himself up to a particular landscape in his experience, to look at it from as many angles as he can, to wonder about it, to dwell upon it.

MOMADAY (1998, 45)

The first Earth Day, April 22, 1970, was conceived as a nation-wide "teach-in" about environmental topics. Twenty million people participated in that event, which somehow organized itself, according to its founders. In the decades that followed, the Environmental Protection Agency was established and a long list of laws were passed to clean the air and water, deal with hazardous waste sites, keep endangered species from going extinct, and expand our national park and wilderness systems. When Earth Day returns each year, it offers an opportunity to focus again on the best ways to live out our lives on this planet (pl. 92).

Plate 92. In 2009 this poster was used to publicize Earth Day activities in Sacramento. "Earth Day, Every Day" is another worthwhile slogan to consider putting into action.

It may seem like a long stretch from earthworms to earthquakes to Earth Day. My hope is that this book can turn that sequence around and point the way from current concerns back toward appreciation of the basic truth that our footsteps should be taken with awareness of the organisms beneath our feet. Bacteria, mycorrhizal fungi, nematodes, earthworms, and springtails are essential to the health of far more prominent and powerful things. As soil ecosystem processes carry on their invisible tasks, the activity of such minuscule creatures leads to viable populations of California Condors, bighorn sheep, and giant sequoias—populations including every endangered species with star power, and the less conspicuous plants and animals that play their roles in the forests and farm fields, and throughout the entire complex of ecosystems across California. We must appreciate that our welfare depends, finally, on the soil—on the land—on the earth.

ONLINE SOIL AND LAND RESOURCES

Agricultural Information and Advice

Orange County Farm Bureau
   http://orange.cfbf.com/
San Joaquin County Farm Bureau
   www.sjfb.org
Lodi Woodbridge Winegrape Commission
   www.lodiwine.com
Community Alliance with Family Farmers
   http://www.caff.org/
UC Sustainable Agriculture Research and Education Program
   www.sarep.ucdavis.edu/
UC Cooperative Extension and Agriculture Experiment Station
   http://ucanr.org/index.cfm
UC Davis, College of Agriculture and Environmental Sciences
   http://caes.ucdavis.edu/
UC Davis Student Farm
   http://studentfarm.ucdavis.edu/
CSU Chico, College of Agriculture
   www.csuchicoag.org/
CSU Fresno, College of Agriculture Science and Technology
   http://cast.csufresno.edu/
Cal Poly San Luis Obispo, College of Agriculture
   http://cagr.calpoly.edu/
Cal Poly Pomona, College of Agriculture
   www.csupomona.edu/~agri/
Santa Rosa Junior College Agriculture/Natural Resources
   www.santarosa.edu/instruction/instructional_departments/
   agriculture/

San Joaquin Delta College
 www.deltacollege.edu/
Butte College
 www.butte.edu/
Ecological Farming Association
 http://eco-farm.org/

Conservation and Ecosystem Restoration and Preservation

Audubon California Landowner Stewardship Program
 www.audubon-ca.org/LSP/Willow_Slough.htm
Audubon California
 www.audubon-ca.org/
The Nature Conservancy, Sacramento River Project and Dye Creek
 Preserve
 http://nature.org/wherewework/northamerica/states/california/
Cache Creek Conservancy
 www.cachecreekconservancy.org/
Solano Land Trust
 www.solanolandtrust.org/

Federal, State, Regional, and Local Land and Soil Conservation Agencies

California Department of Food and Agriculture
 www.cdfa.ca.gov/
Natural Resources Conservation Service, California
 www.ca.nrcs.usda.gov/
Yolo County Resource Conservation District (RCD)
 www.yolorcd.org/
Solano County RCD
 www.solanorcd.org/
Butte County RCD
 http://buttecountyrcd.org/
Tehama County RCD
 www.carcd.org/wisp/tehama/
Glenn County RCD
 www.carcd.org/wisp/glenn/
Southern Sonoma RCD
 www.sonomamarinrcds.org/district-ssc/
San Joaquin RCD
 www.sjcrcd.org/aboutus/index.asp

Westside RCD
   www.carcd.org/wisp/westside/
RCD of Greater San Diego
   www.rcdsandiego.org/
RCD of the Santa Monica Mountains
   www.rcdsmm.org/
South Coast RCD
   www.californiarcandd.org/socorc&d.htm
East Bay Municipal Utilities District
   www.ebmud.com
California Integrated Waste Management Board
   www.ciwmb.ca.gov/landfills/
USFWS San Joaquin National Wildlife Refuge Complex
   www.fws.gov/sanluis/
USFWS Sacramento Valley National Wildlife Refuge Complex
   www.fws.gov/sacramentovalleyrefuges/

Ecological and Carbon Footprint Information and Online Calculators

Ecological footprint online calculator: Redefining Progress
   www.myfootprint.org
Carbon footprint online calculator: Cool California
   www.coolcalifornia.org/article/calculator

REFERENCES

Allen, O. N. 1957. *Experiments in soil bacteriology.* Minneapolis: Burgess.

American Farmland Trust. 2007. Paving paradise: A new perspective on California farmland conversion. www.farmland.org/programs/states/ca/Feature%20Stories/documents/PavingParadise_AmericanFarmlandTrust_Nov07.pdf (accessed January 31, 2010).

Anderson, M. Kat. 2005. *Tending the wild: Native American knowledge and the management of California's natural resources.* Berkeley: University of California Press.

Arnold, Craig Anthony. 2005. Is wet growth smarter than smart growth? The fragmentation and integration of land use and water. *Environmental Law Review* (March 2005):10152–10178.

Asgill, Ladi, ed. 2008. *California Tillage* (newsletter). Modesto, CA: Sustainable Conservation. www.suscon.org/conservationtillage/pdfs/TillageNewsletter-Winter2008.pdf (accessed April 2, 2009).

Audesirk, Teresa, and Gerald Audesirk. 1996. *Biology: Life on Earth.* Upper Saddle River, NJ: Prentice Hall.

Bakker, Elna. 1984. *An island called California.* Berkeley: University of California Press.

Bank of America, California Resources Agency, Greenbelt Alliance, and Low Income Housing Fund. 1995. *Beyond sprawl: New patterns of growth to fit the new California.* San Francisco: Bank of America Corp.

Baldassare, Mark, Dean Bonner, Jennifer Paluch, and Sonja Petek. 2009. *PPIC statewide survey: Californians and population issues.* Public Policy Institute of California. www.ppic.org/main/publication.asp?i=879 (accessed April 15, 2009).

Blank, Steven C. 1998. *The end of agriculture in the American portfolio.* Westport, CT: Quorum Books.

Bolton, Herbert Eugene. 1927. *Fray Juan Crespi, missionary explorer on the Pacific Coast.* Berkeley: University of California Press.

Brewer, William H. 1966. *Up and down California in 1860–1864.* Berkeley: University of California Press.

Buis, Alan. 2010. *NASA data reveal major groundwater loss in California.* Jet Propulsion Laboratory. www.jpl.nasa.gov/news/news.cfm?release=2009-194 (accessed February 3, 2010).

California Department of Conservation. 2006. *Important farmlands in California, 2006.* www.conservation.ca.gov/dlrp/fmmp/products/Pages/FMMP-MapProducts.aspx (accessed May 23, 2009).

California Department of Fish and Game. 2003. *Atlas of the biodiversity of California.* Sacramento: California Resources Agency.

California Department of Parks and Recreation. 2002. *California outdoor recreation plan 2002.* Sacramento: California Department of Parks and Recreation. http://parks.ca.gov/?page_id=23880 (accessed April 15, 2009).

———. 2005. *Outdoor recreation: Parks and recreation trends in California 2005.* Sacramento: California Resources Agency. http://parks.ca.gov/?page_id=23880 (accessed April 15, 2009).

California Park and Recreation Society. 2009. *Market research to support CPRS brand building initiative.* Sacramento: California Park and Recreation Society. www.cprs.org/pdf/Market_Research_Full_Report.pdf (accessed April 15, 2009).

Carle, David. 2003. *Water and the California dream.* San Francisco: Sierra Club Books; Berkeley: University of California Press.

———. 2004. *Introduction to water in California.* Berkeley: University of California Press.

———. 2006. *Introduction to air in California.* Berkeley: University of California Press.

———. 2008. *Introduction to fire in California.* Berkeley: University of California Press.

Central Pacific Land Company. 1924. Lands in California and Nevada. August 1, 1924, map.

Chambers, Nicky, Craig Simmons, and Mathis Wackernagel. 2000. *Sharing nature's interest: Ecological footprints as an indicator of sustainability.* London and Sterling, VA: Earthscan.

Clar, C. Raymond. 1959. *California government and forestry, from Spanish days until the creation of the Department of Natural Resources in 1927.* Sacramento: State of California, Department of Natural Resources, Division of Forestry.

Cleland, Robert Glass. 1941. *The cattle on a thousand hills.* San Marino, CA: Huntington Library.

Cobb, N. A. 1915. Nematodes and their relationships. In *1914 Yearbook of the United States Department of Agriculture*, 457–490. Washington, DC: U.S. Government Printing Office. http://naldr .nal.usda.gov (accessed January 30, 2009).

Coleman, David C., D. A. Crossley Jr., and Paul F. Hendrix. 2004. *Fundamentals of soil ecology*, 2nd ed. Burlington, MA: Elsevier Academic Press.

Collier, Michael. 1999. *A land in motion: California's San Andreas Fault.* San Francisco: Golden Gate National Parks Association; Berkeley: University of California Press.

Cool California. 2009. Carbon calculator. www.coolcalifornia.org/ article/calculator (accessed May 23, 2009).

Darwin, Charles. 1881. *The formation of vegetable mould through the actions of worms with observations of their habits.* London: John Murray. http://darwin-online.org.uk/content/frameset? itemID=F1357&viewtype=side&pageseq=1 (accessed January 30, 2009).

Davis, Mike. 1998. *Ecology of fear: Los Angeles and the imagination of disaster.* New York: Vintage Books.

Diamond, Henry L., and Patrick F. Noonan. 1996. *Land use in America.* Washington, DC: Island Press.

Durrenberger, Robert W. 1968. *Patterns on the land.* Palo Alto, CA: National Press Books.

Evans, Howard Ensign. 1993 [1966]. *Life on a little-known planet: A biologist's view of insects and their world.* New York: Lyons Press.

Farb, Peter. 1959. *Living Earth.* New York: Harper Colophon Books.

Fimrite, Peter. 2008. "Conservation on a staggering scale" at Tejon. *San Francisco Chronicle*, May 9, 2008, A1.

Fisher, Brian L., and Stefan P. Cover. 2007. *Ants of North America.* Berkeley: University of California Press.

Hearst Ranch Conservation Project. 2009. Hearst Ranch, features and surroundings. Map. San Simeon, CA: Hearst Ranch Conservation Project. www.hearstranchconservation.org (accessed May 23, 2009).

Holder, Charles Frederick. 1905. *Life in the open: Sport with rod, gun, horse and hound in southern California.* New York: Putnam's.

Hundley, Norris Jr. 2001. *The great thirst, Californians and water: A history.* Berkeley: University of California Press.

Isherwood, Justin. 2005. *Book of plough: Essays on the virtue of farm, family and the rural life.* Madison: University of Wisconsin Press.

Johnson, Hans P., and Lian Li. 2007. Birthrates in California. *California Counts, Population Trends and Profiles* 9 (2). www .ppic.org/content/pubs/cacounts/CC_1107HJCC.pdf (accessed April 15, 2009).

Jones, Gary. 2005. *Muck and mystery: Nitrogen transport.* www.gary-jones.org/mt/archives/000178.html (accessed December 28, 2009).

Kennedy, Roger G. 2006. *Wildfire and Americans: How to save lives, property, and your tax dollars.* New York: Hill and Wang.

Kroeber, Theodora. 1976 [1961]. *Ishi in two worlds.* Berkeley: University of California Press.

Kruckeberg, Arthur R. 2006. *Introduction to California soils and plants.* Berkeley: University of California Press.

Lee, K. E. 1985. *Earthworms: Their ecology and relationship with soils and land use.* Sydney: Academic Press.

Leopold, Aldo. 1966 [1949]. *A Sand County almanac.* New York: Sierra Club/Ballantine Books.

Logan, William Bryant. 1995. *Dirt: The ecstatic skin of the Earth.* New York: W.W. Norton.

Louv, Richard. 2005. *Last child in the woods: Saving our children from Nature-Deficit Disorder.* Chapel Hill, NC: Algonquin Books.

Lowenfels, Jeff, and Wayne Lewis. 2006. *Teaming with microbes: A gardener's guide to the soil food web.* Portland: Timber Press.

Madison, Mike. 2002. *Walking the flatlands: The rural landscape of the lower Sacramento valley.* Berkeley, CA: Heyday Books.

McPhee, John. 1993. *Assembling California.* New York: Farrar, Straus and Giroux.

Meine, Curt. 2004. *Correction lines: Essays on land, Leopold, and conservation.* Washington, DC: Island Press.

Mitchell, Jeff P., and Gene Miyao. 2002. *Status of conservation tillage research and on-farm demonstrations in California: A 2002 progress report.* Davis: Agriculture and Natural Resources, University of California at Davis. http://groups.ucanr.org/ucct/Status_of _Conservation_Tillage_Research_%26_On-Farm _Demonstrations_in_California/ (accessed May 24, 2009).

Mitchell, Samuel Augustus. 1850. Map of the state of California, the territories of Oregon and Utah, and the chief part of New

Mexico. Philadelphia: Thomas Cowperthwait and Co. www
.davidrumsey.com (accessed May 22, 2009).

Momaday, N. Scott. 1998. *The man made of words: Essays, stories, passages.* New York: St. Martin's Griffin.

Montgomery, David R. 2007. *Dirt: The erosion of civilizations.* Berkeley: University of California Press.

Nardi, James B. 2007. *Life in the soil: A guide for naturalists and gardeners.* Chicago: University of Chicago Press.

NationalAtlas.gov. 2009. *The public land survey system.* http://nationalatlas.gov/articles/boundaries/a_plss.html (accessed May 23, 2009).

Newmark, Harris. 1916. *Sixty years in southern California, 1853–1913.* New York: Knickerbocker Press.

Nordhoff, Charles. 1873. *California: For health, pleasure, and residence.* New York: Harper and Brothers.

Paton, Alan. 1987 [1948]. *Cry, the beloved country.* New York: Scribner.

Pettley, John W. 2009. *The Mount Diablo initial point, its history and use.* Mount Diablo Surveyors Historical Society. www.mdshs.org/article.html (accessed May 23, 2009).

Pimentel, David, and Mario Ciampietro. 1994. *Food, land, population and the U.S. economy.* Washington, DC: Carrying Capacity Network.

Pollan, Michael. 2006. *The omnivore's dilemma: A natural history of four meals.* New York: Penguin Press.

Powell, G. Harold. 1990. *Letters from the Orange Empire.* Los Angeles and Redlands: Historical Society of Southern California.

Redefining Progress. 2009. *How big is your ecological footprint?* www.myfootprint.org (accessed May 23, 2009).

Reisner, Marc. 2003. *A dangerous place: California's unsettling fate.* New York: Pantheon Books.

Robinson, W. W. 1979 [1949]. *Land in California: The story of mission lands, ranchos, squatters, mining claims, railroad grants, land scrip, homesteads.* Berkeley: University of California Press.

Salt, George, F. S. J. Hollick, F. Raw, and M. V. Brian. 1948. The arthropod population of pasture soil. *Journal of Animal Ecology* 17 (2):139–150.

Save Open-Space and Agricultural Resources (SOAR). 2007. www.soarusa.org.

Schaller, Friedrich. 1968. *Soil animals.* Ann Arbor: University of Michigan Press.

Shinn, Charles Howard. 1948 [1885]. *Mining camps. A study in American frontier government.* New York: Alfred A. Knopf.

Smithsonian, National Museum of Natural History. 2009. Dig it! exhibit. http://forces.si.edu/soils/02_03_01.html (accessed May 23, 2009).

Soil and Water Conservation Society. 2000. *Soil biology primer,* rev. ed. Ankeny, IA: Soil and Water Conservation Society.

Starr, Kevin. 1973. *Americans and the California dream, 1850–1915.* New York: Oxford University Press.

———. 2005. *California: A history.* New York: Modern Library.

Steinbeck, John. 1939. *The grapes of wrath.* New York: Viking Press.

Storer, Tracey I., Robert L. Usinger, and David Lukas. 2004. *Sierra Nevada natural history.* Berkeley: University of California Press.

Sullivan, Preston. 2004. *Sustainable soil management.* Davis, CA: National Center for Appropriate Technology. http://attra.ncat .org/attra-pub/soilmgmt.html (accessed March 5, 2009).

Tiedemann, Arthur R., and Carlos F. Lopez. 2004. Assessing soil factors in wildland improvement programs. *In Restoring Western Ranges and Wildlands,* 39–56, USDA Forest Service Gen. Tech. Rep. RMRS-GTR-136.

U.S. Department of Agriculture, Natural Resources Conservation Service. *Soil education lessons—Texture.* http://soils.usda.gov/ education/resources/k_12/lessons/texture/ (accessed May 23, 2009).

U.S. Geological Survey. *Delta subsidence in California: The shrinking heart of the state.* www.baydeltalive.com/assets/application/pdf/ USGS_Subsidence_Brochure.pdf (accessed October 4, 2008).

———. 2005. *Putting down roots in earthquake country: Your handbook for the San Francisco Bay region.* General Information Product 15. Washington, DC: United States Geological Survey. http:// pubs.usgs.gov/gip/2005/15/ (accessed April 12, 2009).

———. 2007. *Riding the storm: Landslide danger in the San Francisco Bay area.* General Information Product 48 (DVD). Washington, DC: United States Geological Survey.

———. 2009. California's parks and public lands. Map. http://education .usgs.gov/california/maps/parks1.htm (accessed May 23, 2009).

U.S. Supreme Court. 1980. *California v. Nevada,* 447 U.S. 125. On exceptions to report of special master. No. 73, orig. argued April 14, 1980. Decided June 10, 1980.

Vail, Mary C. 1888. *Both sides told, or, southern California as it is. . .* Pasadena, CA: West Coast Publishing.

Warren, Earl. 1977. *The memoirs of Chief Justice Earl Warren.* Garden City, NY: Doubleday.

Whitman, Walt. 1919 [1855]. *Leaves of grass.* Garden City, NY: Doubleday, Page, and Co.

Wilson, Edward O. 2010. Within one cubic foot. *National Geographic* 217 (2):62–83.

Winchester, Simon. 2005. *A crack in the edge of the world: America and the Great California Earthquake of 1906.* New York: HarperCollins.

Winsome, Thais, and Paul Hendrix. 2007. *Earthworm ecology in California.* Integrated Hardwood Range Management Program, University of California. http://danr.ucop.edu/ihrmp/oak99.htm (accessed November 16, 2008).

Wolfe, David W. 2001. *Tales from the underground: A natural history of subterranean life.* Cambridge, MA: Perseus.

Wood, Hulton B., and Samuel W. James. 1993. Native and introduced earthworms from selected chaparral, woodland, and riparian zones in southern California. Gen. Tech. Rep. PSW-142. Albany, CA: U.S. Department of Agriculture, Forest Service, Pacific Southwest Research Station.

Woodward, Arthur. 1961. *Camels and surveyors in Death Valley: The Nevada–California border survey of 1861.* Amargosa Valley, NV: Death Valley '49ers.

Woolley, John T., and Gerhard Peters. The American Presidency Project. Santa Barbara, CA. www.presidency.ucsb.edu/ws/?pid= 15373 (accessed March 24, 2010).

ART CREDITS

Plates

ROCHELLE BARCELLONA, Barcellona Inc., www.barcellonabrand.com 92

JUDY BROUGHTON 62, 88

PETER BRYANT 31

DAVID CARLE 1–5, 7–9, 13, 14, 16, 17, 22, 23, 25, 27–30, 32, 37, 38, 43, 47, 48, 52, 51, 57, 61, 64, 69, 70, 72, 79, 80, 83, 85, 89–91

RYAN CARLE 26, 41, 51, author photo

CALIFORNIA DEPARTMENT OF WATER RESOURCES 74, 77, 83

CALIFORNIA STATE PARKS, Fred Andrews, Mendocino District 56; San Luis Obispo District 27

EDDIE DUNBAR 24, 33

PHIL ENSLEY 65

SANTIAGO ESCRUCERIA 35

GREAT VALLEY CENTER 87

JOEL W. HEDGPETH, courtesy of the Phoebe Apperson Hearst Museum of Anthropology and the Regents of the University of California, Catalogue No. 15-20880 42

RICK KATTELMANN 45, 53, 54

ARTHUR R. KRUCKEBERG 11

PETER LATOURRETTE 40

B.F. LOOMIS, courtesy of the National Park Service 71

MIKE LYNCH, courtesy of 55

MYCOHRRIZAL APPLICATIONS, www.mycorrihizae.com 15

NASA, GSFC/METI/ERSDAC/JAROS/AND U.S./JAPAN ASTER SCIENCE TEAM 46, 76

NATIONAL PARK SERVICE, Golden Gate NRA 58

SOIL AND WATER CONSERVATION SOCIETY, Elaine R. Ingham 18

UNIVERSITY OF SOUTHERN CALIFORNIA, DEPARTMENT OF SPECIAL COLLECTIONS 86

USDA AGRICULTURAL RESEARCH SERVICE, Markus Dubach 12; Eric Erbe 19; William Wergin and Richard Sayre 21, 81

USDA NATURAL RESOURCES CONSERVATION SERVICE 75, 81

USDA SOIL AND WATER CONSERVATION SERVICE, Kerry Arroues 10

U.S. FISH & WILDLIFE SERVICE 34; Gary Stolz 39, 60; Joe Skorupa 78

U.S. GEOLOGICAL SURVEY 6, 66, 68, 73; J.K. Nakata 67

BRIAN VALENTINE 20

CHARLES L. WEED, courtesy of the Bureau of Reclamation 44

COLIN WOOLLEY 36

Figures

FIGURE 1, redrawn from Smithsonian, 2009.

FIGURE 2, redrawn from USDA Natural Resources Conservation Service.

FIGURE 3, redrawn from Soil and Water Conservation Society, 2000.

FIGURES 4 AND 5, redrawn from Audesirk, 1996.

FIGURES 6 AND 7, redrawn from Wood and Janet, 1993.

FIGURE 8, redrawn from Storer, Unsinger, and Lukas, 2004.

FIGURE 9, redrawn from Redefining Progress, 2009.

FIGURE 10, redrawn from Cool California, 2009.

Maps

MAP 1, Carle, 2009.

MAPS 2 AND 4, redrawn from Durrenberger, 1968.

MAP 3, from Kruckeberg, 2006.

MAP 5, detail from Mitchell, 1850. David Rumsey Map Collection. www.davidrumsey.com.

MAP 6, redrawn from Nationalatlas.gov, 2009.

MAP 7, redrawn from Pettley, 2009.

MAP 8, detail from Central Pacific Land Company, 1924. Courtesy of Bureau of Land Management.

MAP 9, redrawn from California Department of Conservation, 2006.

MAP 10, redrawn from U.S. Geolgical Survey, 2009.

MAP 11, redrawn from Hearst Ranch Conservation, 2009.

MAP 12, redrawn from Fimrite, 2008.

MAP 13, redrawn from Collier, 1999.

MAPS 14–17, U.S. Geological Survey.

MAP 18, Carle, 2006.

Part Openers

PART 2 Earth Sciences and Image Analysis Laboratory, NASA Johnson Space Center

ADDITIONAL CAPTIONS

HALF-TITLE PAGE, p. i   Near their holes, ants improve soil fertility by depositing "compost heaps" of debris from the foods they have scavenged.

TABLE OF CONTENTS, p. vi   "Footprints on the land" can have two meanings: the direct impacts of walking on the surface, or the less obvious ecological impacts that each Californian has on the planet's resources.

PART OPENER, pp. 70–71   A desert landscape transformed by irrigation: Imperial Valley farms near the Salton Sea, photographed from space in 1991.

PART OPENER, pp. 96–97   Death Valley National Park protects over 3 million acres of desert wildlands in California.

INDEX

ABOUT THE AUTHOR

David Carle received a bachelor's degree from the University of California at Davis in wildlife and fisheries biology and a master's degree from California State University at Sacramento in recreation and parks administration. He was a ranger for the California State Parks for 27 years. He worked at various sites, including the Mendocino Coast, Hearst Castle, the Auburn State Recreation Area, the State Indian Museum in Sacramento, and, from 1982 through 2000, the Mono Lake Tufa State Reserve. He has taught biology and natural history courses at Cerro Coso Community College (the Eastern Sierra College Center) in Mammoth Lakes. David has written several books, among them *Introduction to Fire in California, Introduction to Air in California,* and *Introduction to Water in California* (University of California Press, 2008, 2006, and 2004); *Water and the California Dream: Choices for the New Millenium* (Sierra Club Books, 2003); *Burning Questions: America's Fight with Nature's Fire* (Praeger, 2002); and *Mono Lake Viewpoint* (Artemisia Press, 1992).

Series Design:	Barbara Jellow
Design Enhancements:	Beth Hansen
Design Development:	Jane Tenenbaum
Cartography:	Lohnes & Wright
Composition:	Publication Services, Inc.
Text:	9.5/12 Minion
Display:	Franklin Gothic Typefaces
Printer and Binder:	Golden Cup Printing Company Limited

Regional Guides